Lecture Notes on
Particle Systems and Percolation

THE WADSWORTH & BROOKS/COLE STATISTICS/PROBABILITY SERIES

Series Editors
Ole Barndorff-Nielsen, *Arhus University*
Peter J. Bickel, *University of California, Berkeley*
William S. Cleveland, *AT&T Bell Laboratories*
Richard M. Dudley, *Massachusetts Institute of Technology*

R. Becker, J. Chambers, A. Wilks, *The New S Language: A Programming Environment for Data Analysis and Graphics*
P. Bickel, K. Doksum, J. Hodges, Jr., *A. Festschrift for Erich L. Lehmann*
G. Box, *The Collected Works of George E. P. Box, Volumes I and II*, G. Tiao, editor-in-chief
L. Breiman, J. Friedman, R. Olshen, C. Stone, *Classification and Regression Trees*
J. Chambers, W.S. Cleveland, B. Kleiner, P. Tukey, *Graphical Methods for Data Analysis*
R. Durrett, *Lecture Notes on Particle Systems and Percolation*
F. Graybill, *Matrices with Applications in Statistics, Second Edition*
L. Le Cam, R. Olshen, *Proceedings of the Berkeley Conference in Honor of Jerzy Neyman and Jack Kiefer, Volumes I and II*
P. Lewis, E. Orav, *Simulation Methodology for Statisticians, Operations Analysts, and Engineers*
H. J. Newton, TIMESLAB
J. Rawlings, *Applied Regression Analysis: A Research Tool*
J. Rice, *Mathematical Statistics and Data Analysis*
J. Romano, A. Siegel, *Counterexamples in Probability and Statistics*
J. Tanur, F. Mosteller, W. Kruskal, R. Link, R. Pieters, G. Rising, E. Lehmann, *Statistics: A Guide to the Unknown, Second Edition*
J. Tukey, *The Collected Works of J. W. Tukey*, W. S. Cleveland, editor-in-chief
  Volume I: Time Series: 1949-1964, edited by D. Brillinger
  Volume II: Time Series: 1965-1984, edited by D. Brillinger
  Volume III: Philosophy and Principles of Data Analysis: 1949-1964, edited by L. Jones
  Volume IV: Philosophy and Principles of Data Analysis: 1965-1986, edited by L. Jones
  Volume V: Graphics: 1965-1986, edited by W. S. Cleveland

Lecture Notes on Particle Systems and Percolation

RICHARD DURRETT
Cornell University

W Wadsworth & Brooks/Cole Advanced Books & Software
Pacific Grove, California

Wadsworth & Brooks/Cole Advanced Books & Software
A Division of Wadsworth, Inc.

Printed in the United States of America

10 9 8 7 6 5 4 3 2 1

Library of Congress Cataloging-in-Publication Data

Durrett, Richard, [date]-
     Lecture notes on particle systems and percolation / by Richard
Durrett.
          p.   cm.
     Bibliography: p.
     Includes index.
     ISBN 0-534-09462-7
     1. Percolation (Statistical physics) 2. Mathematical pyhsics.
I. Title.
QC174.85.P45D87 1988
530.1'3--dc 19   88-12160
                                                    CIP

Sponsoring Editor: *John Kimmel*
Editorial Assistant: *Maria Tarantino*
Production Editor: *Sue Ewing*
Manuscript Editor: *Barbara Kimmel*
Art Coordinator: *Lisa Torri*
Printing and Binding: *The Maple-Vail Book Manufacturing Group,
 York, Pennsylvania*

PREFACE

This section explains HOW the book came to be, and WHY it has the form it does. The Introduction explains WHAT it is about.

The idea (or excuse) for writing this book came when I was invited to give some lectures at Anhui Normal University in the People's Republic of China in January of 1986. The aim of those lectures, and of this book, is to take people who know some probability and, in a short amount of time, teach them about some of the results and techniques in the study of interacting particle systems. (To read this book you should know something about Poisson processes, random walks, Markov chains, Chebyshev's (or Markov's) inequality, and Birkhoff's ergodic theorem.)

The six two-hour lectures I gave in China covered the main results in Sections 1, 3, 4a, 5a, 10a, and 10b, and were the embryonic form of the text that follows. Soon after giving the lectures, I had the idea that "I would write a book that explained the ideas behind the proofs like one does in seminar talk; or, even better, when one talks to friends scribbling the main ideas on blackboards or beer-stained pieces of paper."

The outline of the book is true to the dream - most sections correspond to talks I have given, some many times. BUT, to steal a line from Hunter Thompson, "the book is a failed experiment in gonzo journalism." My great idea collapsed under the weight of the realization that you can ask a question in a lecture if there is not enough detail, or skip ahead in a book if there is too much. As a result, the book contains many more details than even the most patient listener would sit still for. The idea is still to explain the ideas, and have some FUN while doing so.

Several of my (former) friends have asked me, "Why are you writing a book on particle systems? Tom Liggett has already done that." The frivolous answer is - while Tom's book has the meat of the subject, this book also has the cheese, lettuce, onion, mustard, and sesame seed bun needed to make a cheeseburger. More seriously, the books are orthogonal in content and style. Liggett's topic is particle systems. The unifying theme here is percolation. We consider only the particle systems that can be constructed from graphical representations, and then we study percolation (oriented and regular varieties), and first passage percolation. Even where the books overlap (the voter model and the contact process), the treatments are different. Throughout our book, theorems are illustrated by computer simulations and the discussion is littered with jokes. We prove some very recent results on the voter model and discuss asymptotic shape results, which Liggett does not treat. Of course, there are a few things in Liggett's book that we do not consider here: the Ising model, nearest-particle systems, the exclusion process, and "linear systems."

To wrap up these ramblings, we would like to explain two of the book's shortcomings. (I) *It is not a history book.* We content ourselves to give the best proofs we know and refer the reader to the literature for the missing details. This robs some of the early contributors of the credit they deserve. To rectify this, we have added a set of notes at the end of the book to help set the record straight. (II) *It is a history book.* We only do things that have reached a fairly definitive form. Consequently, the reader will find no mention of percolation in $d \geq 3$, or hydrodynamic limits or large deviations for particle systems, and very little discussion of correlation lengths and critical exponents, even though these are all very active topics. We do not totally ignore the future though - about twenty (very hard) open problems are mentioned in the text, and the notes referred to above indicate where the reader can find out about recent developments.

Finally, some thank you's are in order. Gold medals go to Ted Cox and Glen Swindle who carefully read the entire manuscript and pointed out hundreds of errors. Silver medals go to David Griffeath and Tom Liggett who read large parts of the manuscript and made valuable comments on its weaknesses; and to Xiaolong Lou who showed me many places where a newcomer to the subject could have trouble. Honorable mentions go to Florin Avram, Ding wan-ding, Harry Kesten, and Roberto Schonmann who each pointed out several mistakes.

Next, I should thank my colleagues and coauthors, without whom this book would not exist: Maury Bramson, Lincoln and Jennifer Chayes, Ted Cox, Larry Gray, David Griffeath, Harry Kesten, and Roberto Schonmann. Special thanks go to Tom Liggett for teaching me the subject, and to Ted Harris for inventing the contact process and graphical representation.

Before I get too far down the list, I have to thank my wife Susan. Despite momentary lapses ("Why are you writing a book when all the results have appeared in papers?"), she was supportive and patient during the project. Last but not least, I am grateful for the financial support of the National Science Foundation and the Army Research Office (through the Mathematical Sciences Institute at Cornell University). If you would like to see your foundation listed here (or in my next book), send your check to

Rick Durrett
Department of Math/White Hall
Cornell University
Ithaca, NY 14853

CONTENTS

INTRODUCTION

In this section, we give a "cocktail party" summary of the book. We describe the processes we will consider and state the main results. Hopefully, this will help keep the reader from getting lost in the details and allow the reader to skip ahead if nausea sets in. To keep the introduction to a suitable length, we have kept explanations to a minimum. Everything will be fully explained when the time comes. Our story is divided into four acts.

ACT I (Chapters 1–5):

Particle Systems Constructed from Graphical Representations. In the four processes considered in this segment, the state at time t, $\xi_t \subset \mathbb{Z}^d$, and ξ_t is the set of "occupied sites". Points are added or deleted from ξ_t at rates that depend upon $n_t(x)$ = the number of occupied neighbors, the neighbors of a point in $\mathbb{Z}^d$ being the 2d points closest to x.

If $x \notin \xi_t$ then $P(x \in \xi_{t+s} \mid \xi_t) = s\,\beta(n_t(x)) + o(s)$.

If $x \in \xi_t$ then $P(x \notin \xi_{t+s} \mid \xi_t) = s\,\delta(n_t(x)) + o(s)$.

The birth rates β and death rates δ depend upon the model.

In Chapter 1, we consider Richardson's model, the process with $\beta(k) = k$ and $\delta(k) = 0$. Here sites become occupied at a rate equal to the number of occcupied neighbors, and once occupied, never become vacant. In this process, ξ_t increases to cover all of $\mathbb{Z}^d$, and attention focuses on how the set grows. The main result shows that if ξ_0 is finite, then the diameter of ξ_t grows linearly, and ξ_t has an asymptotic shape.

In Chapter 2, we introduce the voter model, the process with $\beta(k) = k$ and $\delta(k)$ $= 2d - k$. The birth rate is the same as before, but now sites become vacant at a rate proportional to the number of vacant neighbors. The reason for the name becomes apparent if we imagine that each point is occupied by a Democratic or Republican voter, and the voter changes her party affiliation at a rate equal to the number of neighbors in that party. Here, ξ_t is the set of voters in one party, and we follow Spock and Rothenberg (1985) in using the feminine pronoun, except on occasions when we forget about our convention.

If $\mathbb{Z}^d$ were a finite set, the asymptotic behavior of the voter model would be trivial. Eventually all the sites would have the same opinion, and the state would not change. However, $\mathbb{Z}^d$ is infinite, so the answer there is more interesting. In $d \leq 2$, the system approaches complete consensus starting from any initial configuration. In $d > 2$, if we start from product measure with density θ, then the system approaches a nontrivial stationary distribution (which is different for each $\theta \in [0,1]$). (Note: after reading Chapter 2, the reader is mathematically, but probably not psychologically, ready for Chapter 10 where recent results on the voter model are described.)

Chapter 3 concerns the biased voter model, the process with $\beta(k) = \lambda k$ and $\delta(k)$ $= 2d - k$. In this crude model of carcinogenesis, ξ_t is the set of infected sites. Infected sites become healthy at a rate equal to the number of healthy neighbors, and healthy sites become infected at rate λ times the number of infected neighbors. An easy calculation shows that if $\lambda > 1$, and the system starts with one infected site, then with probability $(\lambda-1)/\lambda$, the tumor does not die out. The main result asserts that when the tumor does not die out, it grows linearly and has an asymptotic shape.

Chapter 4 is the first of two chapters on the contact process. This model has the same birth rate as the biased voter model but a simpler looking death rate, $\delta(k) \equiv 1$. In this model, there is a critical value λ_c so that if $\lambda > \lambda_c$ the infection has a positive probability of not dying out starting from $\xi_0 \neq \phi$, and if $\lambda < \lambda_c$ the system dies out whenever ξ_0 is finite. In the biased voter model, $\lambda_c = 1$. In the contact process, $\lambda_c \approx$ 1.65 but the exact value of λ_c is not known.

The main result of Section 4a is a characterization of λ_c for the one–dimensional model, which allows us to prove results for all $\lambda < \lambda_c$ without knowing what λ_c is. This characterization is applied in Section 4b to prove two result about the behavior starting from a finite set. If $\lambda < \lambda_c$ then the process dies out exponentially fast. If $\lambda > \lambda_c$ then on the set of nonextinction, ξ_t grows linearly.

ACT II (Chapter 5):

Discrete Time Models. In this segment, we introduce the discrete cousins of the characters introduced in the first part and set the stage for the change of plot in act three. In Section 5a, we introduce oriented percolation. This model takes place on $V = \{(m,n) : m,n \in \mathbb{Z} \text{ and } m+n \text{ is even}\}$. There is an oriented bond from (m,n) to (m−1,n+1) and to (m+1,n+1), and these bonds are independently open or closed with probabilities p and 1−p. Using this "percolation structure," we can define a process by $\xi_n^A = \{ y : (x,0) \to (y,n) \text{ for some } x \in A \}$, where a → b means there is a path of open bonds from a to b. This process is the discrete time analogue of the contact process.

In Section 5b, we enlarge the class of discrete time models we can construct by allowing the state of the bonds that end at the same site to be dependent. These models have properties very similar to the contact process. In Section 5c, we go "back to continuous time" to show that the graphical representations that we used in Chapters 1–4 to construct the processes in part one are limits of structures like the ones used in Section 5b. In the last two–fifths of Chapter 5, we consider "Pascal's triangle mod 2" and other "cancellative systems" which are constructed by setting η_n^A

= { y : (y,n) can be reached from A × {0} by an odd number of paths}. This little change in the definition (i.e., replacing "at least one" by "an odd number of") makes a big difference in the behavior of the models, and there are many open problems.

ACT III (Chapters 6–9):

Percolation and First Passage Percolation. The third act of a play traditionally occurs after intermission, and, in our case, it offers the reader a chance to begin afresh. Chapters 6–9 are indpendent of Chapters 2–5.

In Chapter 6, we consider percolation processes on $\mathbb{Z}^2$. In the bond version of this model, discussed in Section 6a, there is an arc (or bond) connecting each pair of neighbors in $\mathbb{Z}^2$, and each bond is independently open with probability p and closed with probability 1–p. Open bonds are thought of as air spaces, and attention focuses on C_0 = the set of points that can be reached from 0 by a path of open bonds. When P(C_0 is infinite) > 0, we say that percolation occurs. Most of Section 6a is devoted to a proof of Kesten's result that the critical value p_c = inf{ p : P(C_0 is infinite) > 0 } is 1/2. The ideas in this proof are the keys to the developments in Chapters 7–9.

The second half of Chapter 6 is devoted to site percolation. This model is like bond percolation, but it is the sites x ∈ $\mathbb{Z}^2$ that are designated as open or closed. Much of the theory is the same but this time the critical value is not known. In Chapter 7, we make an excursion into the realm of (random) fractals to study a percolation process introduced by Mandelbrot. The main reason for our interest in this model is that it has a "first–order phase transition" – there is positive probability of percolation at the critical value.

In Chapter 8, we turn our attention to first passage percolation in $\mathbb{Z}^2$. In this model, each bond has an independent nonnegative random variable attached to it which is the amount of time it takes the fluid to traverse the bond. With this interpretation in mind, we let t(0,x) = the first time the fluid will appear at x if there is a source at 0 that begins operating at time 0. If we let $\xi_t = \{ x : t(0,x) \le t \}$ then the main result of Section 8a gives necessary and sufficient conditions for ξ_t to have an asymptotic shape.

In Section 8b, we consider the special case in which the passage times are 0 and 1 with probabilities p and 1–p, and focus our attention on the "time constant" $\mu(p) =$ lim t(0,(n,0))/n. As the reader can probably guess, this special case is closely related to bond percolation. Our results show that $\mu(p) > 0$ if and only if p < 1/2, and t(0,(n,0)) ≈ log n when p = 1/2.

In Chapter 9, we investigate some processes that have been used to model epidemics and forest fires. Adopting the terminology of the second application, there is a tree at each point x ∈ $\mathbb{Z}^2$ that is healthy, on fire, or burnt. Healthy trees catch fire at rate λ times the number of burning neighbors, burning trees stay on fire for a random amount of time with distribution F, and burnt trees stay burnt. This model can be studied by a combination of the techniques from Sections 6a and 8a. The main result asserts that starting from a single burning tree in the middle of an otherwise healthy forest, then, when the fire does not go out, the set of burnt trees grows linearly and has an asymptotic shape, and all of the fire is near the boundary of the growing ball.

ACT IV: (Chapters 10–11):

Particle Systems, II. The tone of the play changes in the final act. Most of the material in the last two chapters is recent work and/or requires a substantial amount of effort to carry out the details. We concentrate mostly, then, on explaining the ideas that underlie the proofs.

In Chapter 10, which is independent of Chapters 2–9, we describe some recent results on the voter model. As we mentioned above, the voter model approaches complete consensus in d $\leq$ 2. In Section 10a, we take a close look at how this occurs in d = 1 (which is easy), and in d = 2 (where the answer, due to Cox and Griffeath, is intricate and beautiful). In Section 10b, we look at the voter model on the torus $(\mathbb{Z} \bmod N)^d$. That process is trivial in the sense that at some time τ all the opinions will become the same, and the state will not change after time τ. The main result, due to Ted Cox, identifies how τ grows with N. The answer depends upon the dimension d.

In Chapter 11 we return to the contact process. This chapter is independent of Chapters 7–10, but uses ideas from Sections 5a and 6a in its first section, where the "renormalized bond construction" is described. This construction, which is almost as ugly as it is powerful, is the key to the developments that follow. In Section 11b, we prove results that show that "everything happens exponentially fast when $\lambda > \lambda_c$", and we use these results to prove a law of large numbers for $|\xi_t^0|/t$.

Complete proofs are given for the main results of Sections 11a and 11b, but then the level of detail goes downhill fast. In Section 11c, we introduce a general asymptotic shape result and sketch its proof. In Section 11d, we apply this result to complete the proof of the shape theorem for the biased voter model from Chapter 3, and to prove a similar result for the contact process. To be honest, we start to do this, then get bored with the details, and change the subject to talk about what is known and what we would like to know.

ADVERTISEMENT:

Throughout the book the reader will see pictures of computer simulations. The Pascal programs that generated these pictures (and more) can be found on a disk called IPSmovies (Version 2.0), which is also available from Wadsworth. It is a "computer workbook" for this text with lots of things to see and do.

EPILOGUE:

We close the introduction with "a speech, short poem, or the like, addressed to the spectators serving to explain the plan of the work." There are two things that tie the four acts together.

1. Everything we do is percolation. This should be clear for the material in Acts II and III. In Section 5c, we explain that the graphical representations used to construct the particle systems studied in Acts I and IV are limits of percolation processes.

2. Almost all of the models considered have a critical value and have two properties; when subcritical, they die out exponentially fast, and when supercritical, they grow linearly and have an asymptotic shape. This is true for the biased voter model, contact process, percolation (oriented and regular), and our forest fire. It is also true for Richardson's model and first passage percolation if we consider these models to be always supercritical. Finally, the voter model fits into this picture as the biased voter model at the critical value.

1 The Simplest Growth Models

In these notes we will chiefly be concerned with processes whose states at time t, $\xi_t \subset \mathbb{Z}^d$ = the d–dimensional integer lattice. The points in ξ_t are typically thought of as being occupied by a "particle" or are occasionally called "infected" when we are thinking of the spread of a pest through an orchard of trees. In this section we will confine our attention to the simplest growth models, those with no deaths: if $x \in \xi_s$ then $x \in \xi_t$ for all $t \geq s$. We will start with what is perhaps the simplest example – Richardson's model, the process in which particles are born at a rate proportional to the number of occupied neighbors. Explaining the last sentence will take a few moments. First the neighbors of x, $N(x) = \{y : |x-y| = 1\}$ where $|x-y| = |x_1-y_1| + \ldots + |x_d - y_d|$ is the norm we will use throughout these notes. If we let $|A| = $ the number of points in A and use "$f(h) \sim g(h)$ as $h \to 0$" to denote $f(h)/g(h) \to 1$ then we can formulate the assumption above as:

if $x \notin \xi_t$ then $P(x \in \xi_{t+h} | \mathscr{F}_t) \sim h|\xi_t \cap N(x)|$ as $h \to 0$.

Here, as is usual in discussing Markov processes, $\mathscr{F}_t$ is the σ–field generated by ξ_s, $s \leq t$.

If the discussion in the last paragraph reminds the reader of her training in Markov chains, that is good, but it is not necessary for the developments that follow. We will build all of our processes by hand. For each x and y with $|x-y| = 1$, let $T_n^{(x,y)}$, $n = 1,2,\ldots$ be a Poisson process with rate 1. That is, if we set $T_0^{(x,y)} = 0$ then

(i) $T_n^{(x,y)} - T_{n-1}^{(x,y)}$, $n = 1,2,...$ are independent, and

(ii) $P(T_n^{(x,y)} - T_{n-1}^{(x,y)} \geq t) = e^{-t}$.

At times $T_n^{(x,y)}$ we draw an arrow from x to y to indicate that if x is occupied at that time, then y will become occupied (if it is not already). See Figure 1.1 for a picture.

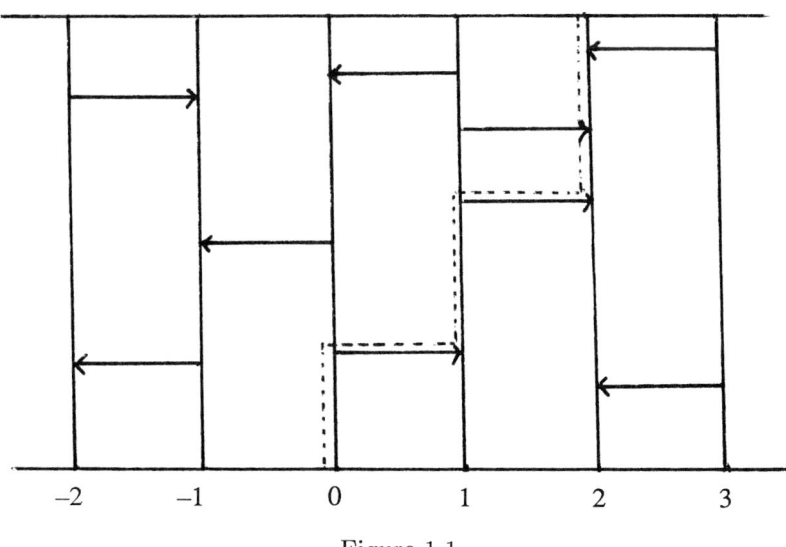

Figure 1.1

Let ξ_t^A denote the state at time t of the process with $\xi_0^A = A$. We define ξ_t^A to be the set of points y such that (y,t) can be reached from some point (x,0) with $x \in A$ by a path that only goes upward or moves sideways in the directions of the arrows. If $A = \{0\}$ in Figure 1.1 then $\xi_t^A = \{-1,0,1,2\}$. We have indicated one path from (0,0) to (2,t) by a dotted line. Notice that we cannot go to (3,t) because the arrow points in the wrong direction.

We will leave it to the reader to check that the process we constructed has the transition rates described above. We turn now to the problem of analyzing the asymptotic behavior of ξ_t^0, the process starting from $\xi_0^0 = \{0\}$. We begin with the trivial case $d = 1$. Let $t(n) = \inf\{\, t : n \in \xi_t^0 \,\}$ or in words, $t(n) = $ the first time n is infected. If $n > 0$ the infection must get to n by spreading first to 1, then to 2, and so on. The lack of memory property of the exponential distribution implies that the successive infections take independent exponentially distributed amounts of time, so the strong law of large numbers implies $t(n)/n \to Et(1) = 1$.

In $d > 1$ there is more than one way for the infection to spread from 0 to x, and things become more subtle. Richardson (1973) was the first to figure out how to deal with this problem. His idea, roughly, was that while the infection times were not additive they were "subadditive," and this is enough to prove a law of large numbers. To formulate this precisely requires some notation. Let $\xi_t^{(x,s)}$ be the set of all points at time t that can be reached from the point x at time s using the same rules as before. One can go up or move sideways in the direction of the arrows. Let $t(x) = \inf\{\, t : x \in \xi_t^0 \,\}$, and $t(x,y) = \inf\{\, u : y \in \xi_{t(x)+u}^{(x,t(x))} \,\}$. In words, $t(x,y)$ is the amount of time it takes the infection to get to y if we start with x infected at time $t(x)$. Since x is infected at time $t(x)$, it is clear that

$$\xi_{t(x)+u}^{(x,t(x))} \subset \xi_{t(x)+u}^0,$$

and hence $t(x) + t(x,y) \geq t(y)$.

By properties of the Poisson process, $t(x,y)$ is independent of $t(x)$ and has the same distribution as $t(y-x)$. If we fix some $x \in \mathbb{Z}^d$ and let $s_{m,n} = t(mx,nx)$, then

(1) $$s_{0,m} + s_{m,n} \geq s_{0,n}.$$

From this "subadditivity" property and the independence mentioned above, it follows easily that we have

(2)
$$s_{0,n}/n \to \inf_{m\geq 1} E\, s_{0,m}/m \text{ in } L^1.$$

To prove (2), we begin by taking expected values in (1), and letting $a_n = E s_{0,n}$ to get

(3)
$$a_m + a_{n-m} \geq a_n.$$

From this it follows easily that

(4)
$$a_n/n \to \inf_{m\geq 1} a_m/m \equiv \mu.$$

To prove (4), we observe that the liminf is clearly $\geq \mu$, so all we have to do is prove that for any m the limsup $\leq a_m/m$. The last fact is easy, for if we write $n = km + \ell$ with $0 \leq \ell < m$ and use (3), then we get $a_n \leq ka_m + a_\ell$. Dividing by n gives

$$\frac{a_n}{n} \leq \left(\frac{km}{km+\ell}\right)\frac{a_m}{m} + \frac{a_\ell}{n}.$$

Letting $n \to \infty$ now, and doing a little arithmetic gives (4).

Repeating the last proof with random variables, we can write

$$\frac{s_{0,n}}{n} \leq \frac{k}{km+\ell} \cdot \frac{s_{0,m} + \cdots + s_{(k-1)m,km}}{k} + \frac{s_{km,n}}{n}.$$

The strong law of large numbers implies that the first term on the right $\to Es_{0,m}/m$ as $n \to \infty$. (Recall m is fixed and $\ell \leq m$.) $Es_{0,\ell} < \infty$ implies that for $\epsilon > 0$

$$\sum_{n=1}^{\infty} P(s_{0,\ell} > \epsilon n) \leq \int_0^{\infty} P(s_{0,\ell} > \epsilon t) \, dt = E(s_{0,\ell}/\epsilon) < \infty.$$

So the Borel–Cantelli lemma implies the other term $\to 0$. Since m is arbitrary, the argument above shows

$$\lim_{n \to \infty} \sup s_{0,n}/n \leq \mu \equiv \inf_{m \geq 1} E s_{0,m}/m.$$

We are now ready for the knockout punch: $s_{0,n}/n$ is a sequence of random variables with limsup $\leq \mu$ and expected values $\to \mu$. If we observe $|x| = 2x^+ - x$, where $x^+ = \max(x,0)$, and write

$$E|s_{0,n}/n - \mu| = 2E(s_{0,n}/n - \mu)^+ - E(s_{0,n}/n - \mu),$$

then the last term $\to 0$, and $(s_{0,n}/n - \mu)^+ \to 0$ a.s. To finish the proof then, it suffices to show that $E(s_{0,n}/n - \mu)^+ \to 0$.

This is a technicality, but one that we will need twice more below, so it pays to do it correctly now. Let $k = |x|$ and let $x_0, x_1, \ldots, x_k$ be a sequence of points with $x_0 = 0$, $x_k = x$, and $|x_i - x_{i-1}| = 1$ for $i = 1, \ldots, k$. Let $t_1 = T_1^{(0,x(1))}$ be the first point in the Poisson process of infections from 0 to x_1, and for $0 \leq j \leq k$ let t_j be the first point $> t_{j-1}$ in the Poisson process $\{T_n^{(x(j-1),x(j))} \ n \geq 1\}$. [Here and in what follows when subscripted superscripts threaten to occur, we will avoid them by writing $x(j)$ for x_j.] It is easy to see that $t(x) \leq t_k$. The lack of memory property of the exponential distribution implies that t_k is a sum of k independent exponential random variables.

The last observation allows us to prove a powerful upper bounds for $t(x)$ (see (8) below) but all we will use now is $Et(x)^2 < \infty$. Since $s_{0,1}, s_{1,2}, ..., s_{n-1,n}$ are independent and have the same distribution as $t(x)$, we have

$$Es_{0,n}^2 \leq (nEs_{0,1})^2 + n \text{ Variance}(s_{0,1}) \leq Cn^2,$$

and hence $E(s_{0,n}/n)^2 \leq C$. Combining this observation with the trivial bound $(s_{0,n}/n-\mu)^+ \leq s_{0,n}/n$, gives more than enough to conclude $E(s_{0,n}/n-\mu)^+ \to 0$ and we are done.

What we have given above is the easy part of the "subadditive ergodic theorem." By working harder, it is possible to extract much more from the subadditivity property. Since Liggett (1985a) has a clear proof to which we have nothing to add, we will quote the result without proof.

(5) Theorem. Suppose $s_{m,n}$, $m \leq n$ are random variables that have the following properties:

(a) $s_{0,0} = 0$, $s_{0,n} \leq s_{0,m} + s_{m,n}$ for $0 \leq m \leq n$;
(b) $\{s_{(n-1)k,nk}, n \geq 1\}$ is a stationary sequence for each k;
(c) $\{s_{m,m+k}, k \geq 0\} = \{s_{m+1,m+k+1}, k \geq 0\}$ in distribution for each m;
(d) $Es_{0,1}^+ < \infty$.

Let $\alpha_n = Es_{0,n} < \infty$, which is well defined by (a), (b) and (d). Then

$$\alpha \equiv \lim_{n \to \infty} \alpha_n/n = \inf_{m \geq 1} \alpha_m/m$$

$$X \equiv \lim_{n \to \infty} s_{0,n}/n \text{ exist a.s., and } EX = \alpha.$$

If the stationary processes in (b) are ergodic then $X = \alpha$ a.s.

Note 1: Some people may have expected us to refer to Kingman's subadditive ergodic theorem. That result requires $s_{\ell,m} + s_{m,n} \geq s_{\ell,n}$ for all $0 \leq \ell < m < n$ (which is false in this application).

Note 2: If you have never seen a subadditive ergodic theorem before, you should note that if $X_1, X_2, \ldots$ is a stationary sequence with $E|X| < \infty$ then $s_{\ell,m} = X_{\ell+1} + \ldots + X_m$ satisfies (a) $-$ (d) with equality in (a), so the last result is an extension of the ergodic theorem. We will see several times below that it is a useful extension. For an immediate example, let S_n be a random walk in $\mathbb{Z}^d$, let $s_{\ell,m} = |\{ S_k : \ell < k \leq m \}|$ be the number of points visited in $(\ell,m]$, and use (5) to prove that

$$|\{ S_m : 0 < m \leq n \}| \, / n \to P(S_m \neq 0 \text{ for all } m \,).$$

Returning to Richardson's model and observing that $s_{0,n} = t(0,nx)$ satisfies the hypotheses of (5), we see that $t(0,nx)/n \to$ a limit $\mu(x)$, so ξ_t^0 grows linearly in each direction. Our next goal is to say something about the limiting shape of ξ_t^0. To do this, it is convenient to fatten up ξ_t^0 by replacing each $x \in \xi_t^0$ by a square of side 1 centered at x. In symbols, we let

$$\bar{\xi}_t^0 = \{ x + y : x \in \xi_t^0, y \in [-\tfrac{1}{2},\tfrac{1}{2}]^d \}.$$

With this notation introduced we can state our first "shape result".

(6) Theorem. There is a convex set A so that for any $\epsilon > 0$

$$P((1-\epsilon)tA \subset \bar{\xi}_t^0 \subset (1+\epsilon)tA) \to 1 \quad \text{as} \quad t \to \infty.$$

Proof: The first step in proving this result is to figure out what A should be. First we extend $t(x)$ to $x \in \mathbb{R}^d$ by setting $t(x) = \inf\{ t : x \in \bar{\xi}_t^0 \}$. If x is such that $mx \in \mathbb{Z}^d$ for some m, it is straightforward to apply (5) to conclude that

(7)   as $n \to \infty, \quad t(nx)/n \to \mu(x)$ a.s.

The bound at the end of the proof of (2) shows that $|t(x)-t(x+y)| \leq$ the sum of $|y|$ mean 1 exponentials, so a routine argument (left to the reader) shows (7) holds for all x.

Having extended the convergence result to all $x \in \mathbb{R}^d$, we can now write down $A = \{ x : \mu(x) \leq 1 \}$. It is easy to see that $\mu(cx) = c\mu(x)$ for $c > 0$ and $\mu(x+y) \leq \mu(x) + \mu(y)$. We will show in a minute that $\mu(x) \neq 0$ for $x \neq 0$, so μ defines a norm on $\mathbb{R}^d$, and A = the unit ball in that norm.

The keys to the proof of the shape result are two exponential estimates.

(8) Lemma. If $\delta < 1$, there are constants $C, \gamma \in (0,\infty)$ so that

$$P(t(x) > |x|/\delta) \leq Ce^{-\gamma|x|}.$$

Proof: As we remarked at the end of the proof of (2), if $|x| = k$, then $t(x) \leq t_k =$ the sum of k independent exponentials, so it suffices to show the result for t_k. The proof is easy, but will be useful many times below, so we spell out the details. If $\theta > 0$ and $\theta < 1$

$$e^{\theta ak} P(t_k > ak) \leq E\exp(\theta t_k) = (1-\theta)^{-k},$$

so

$$P(t_k > ak) \le \exp(f_a(\theta)k),$$

where

$$f_a(\theta) = -\theta a - \log(1-\theta).$$

Now $f_a(0) = 0$ and $f'_a(0) = -a + 1$, so if $a > 1$, we can pick θ small and get $f_a(\theta) < 0$. This proves the result.

(9) Lemma. There are constants $c, C, \gamma \in (0,\infty)$ so that

$$P(\, t(x) < c|x| \,) \le Ce^{-\gamma t}.$$

Proof: Let $n = |x|$ If $t(x) < cn$, there must be a "chain of infection" that reaches x by time n. To describe what we mean by a chain of infection, let $x_0, x_1, \ldots, x_m$ be a sequence of distinct points with $x_0 = 0$, $x_m = x$, and $|x_i - x_{i-1}| = 1$ for $i = 1, \ldots, m$. Let $t_1 = T_1^{(0, x(1))}$ be the first point in the Poisson process of infections from 0 to x_1, and for $0 \le j \le k$ let t_j be the first point $> t_{j-1}$ in the Poisson process $\{T_n^{(x(j-1), x(j))} \; n \ge 1\}$. The lack of memory property of the exponential distribution implies that t_m is a sum of m independent exponential random variables. It is easy to see that $t(x) \le t_m$ and $t(x)$ is the minimum infection time over all such sequences. If $\theta > 0$ then

$$e^{-\theta a m} P(t_m < am) \le E\exp(-\theta t_m) = (1+\theta)^{-m},$$

so

$$P(t_m > am) \le \exp(f_a(\theta)m),$$

where

$$f_a(\theta) = \theta a - \log(1+\theta).$$

For fixed a, the best choice of θ is $\theta_a = (1/a) - 1$, and in this case

$$f_a(\theta_a) = 1 - a - \log(1/a).$$

The number of paths of length m is $\leq (2d)^m$, so

$$P(\, t(x) \leq a|x|\,) \leq \sum_{m=n}^{\infty} (2d)^m \exp(f_a(\theta_a)m),$$

and the desired result follows from this.

Turning to the proof of (6), if we pick $\delta < 1$ then (8) implies

$$\sum_x P(t(x) > |x|/\delta) < \infty.$$

So it follows from the Borel–Cantelli lemma that for only finitely many x, $t(x) > |x|/\delta$ and hence, if t is chosen large enough,

(10) $$\{x : |x| \leq t\delta\} \subset \{\,x : t(x) \leq t\,\} = \bar{\xi}_t^0.$$

At this point we have shown that $\bar{\xi}_t^0$ can cover a small ball. This may not seem like much, but at this point a routine argument takes over and does the rest. For ease of reference, we give this block of text a name.

Denouement. First we prove $P(\bar{\xi}_t^0 \supset (1-\epsilon)tA) \to 1$. If $x \in (1-\epsilon)A$, then (7) implies that with high probability $tx \in \bar{\xi}^0(t(1-\epsilon/2))$. Let $D(y,r) = \{\,x : |x-y| < r\}$. (D stands for diamond. Recall we use the L^1 norm.) Using the observation

$$\xi_{t(x)+u}^{(x,t(x))} \subset \xi_{t(x)+u'}^0$$

and (10), it follows that with high probability $\bar{\xi}_t^0 \supset D(tx,t\delta\epsilon/2)$. The open sets $D(x,\delta\epsilon/2)$ with $x \in (1-\epsilon)A$, cover the compact set $(1-\epsilon)A$, so there is a finite subcover $D(x_i,\delta\epsilon/2)$, $i = 1,...,I$. Since

$$P(\bar{\xi}_t^0 \supset D(tx_i,t\delta\epsilon/2) \text{ for } i = 1,...,I) \to 1$$

as $t \to \infty$, it follows that $P(\bar{\xi}_t^0 \supset t(1-\epsilon)A) \to 1$.

Somewhat remarkably the argument that we just used to "fill up the inside" can be used to "clean off the outside"; i.e., prove that $P(\bar{\xi}_t^0 \subset (1+\epsilon)tA) \to 1$. To do this we make two observations.

(a) For any $x_1,...,x_k$ in A^c, $P(sx_i \in \bar{\xi}_s^0$ for some $1 \leq i \leq k) \to 0$.

(b) If $\bar{\xi}_t^0$ contains a point x outside $t(1+\epsilon)A$ at time t, then it contains a point $y \in Z^d$ with $t(1+\epsilon) \leq |y| \leq t(1+\epsilon)+1$. Using (10) again, we conclude that with high probability $\bar{\xi}^0(t(1+\epsilon/2)) \supset D(y,t\epsilon\delta/2)$.

To obtain the desired conclusion now, we pick $x_1,...,x_k$ in $2A-A$ so that every point in $2A-A$ is within $\epsilon\delta/2$ of some x_i, and let $s = (1+\epsilon/2)t$ in (a). To prepare for the future, we would like the reader to observe that in the denouement we only used (7) and (10).

The shape theorem given above holds for a wide class of models with no deaths. (See Schürger (1979).) One version, which will be of interest in what follows, is the discrete time process $\xi_n \subset Z^d$ which evolves according to the following rules:

(i) if $x \in \xi_n$ then $x \in \xi_{n+1}$, and

(ii) if $x \notin \xi_n$ then $P(\, x \notin \xi_{n+1} \mid \xi_n \,) = (1-p)^{| \xi_n \cap N(x) |}$.

In words, each infected neighbor of x will independently infect x with probability p. An almost word for word repetition of the arguments above shows that (6) holds for this model. The limit set A is clearly $\subset \{ \, x : |x| \leq 1 \, \}$. In Section 5a we will show that A intersects $\{ \, x : x_1 + x_2 = 1 \, \}$ in a nonempty interval when p is close to 1. This is the "flat edge" result of Durrett and Liggett (1981). Almost nothing else is known about the limit set in discrete time.

To give the reader a feel for what the limit set looks like in continuous time, we close this section by giving some computer simulations. To save ink, only points in ξ_t^0 which are on the boundary are colored black. It has been conjectured that the limiting shape is a circle in this case. Can the reader prove or disprove this?

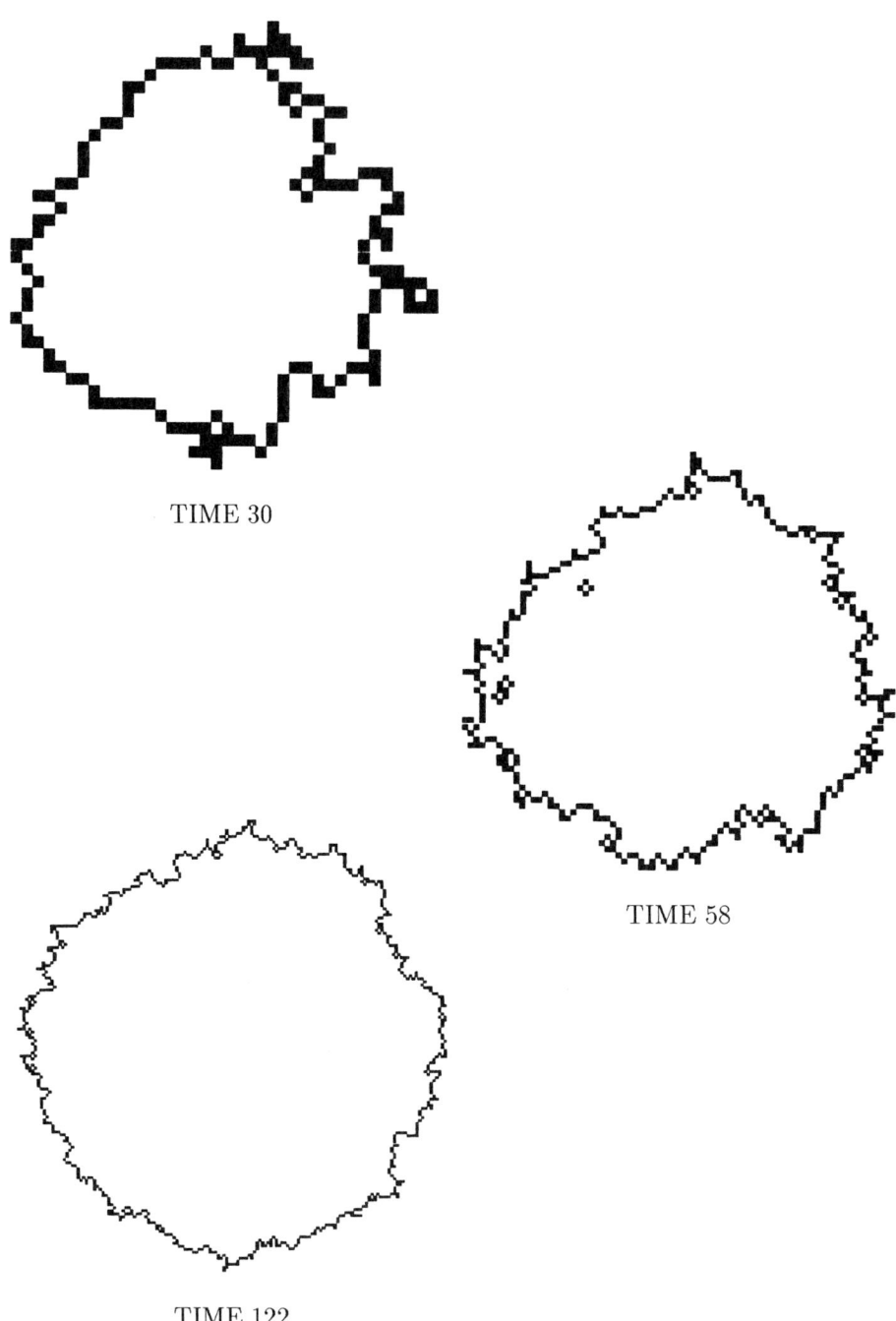

TIME 30

TIME 58

TIME 122

2 The Voter Model

In this chapter we will consider the process that we judge to be the second simplest example of an interacting particle system. In the voter model we think of the sites of $\mathbb{Z}^d$ as being occupied by persons who are either in favor of or opposed to some issue. To write this as a set–valued process, we let ξ_t = the set of voters in favor. Our voters are very simple minded, and change their opinion at a rate equal to $1/2d$ times the number of neighbors who have the opposite opinion.

To construct the voter model, let $\{T_n^x, n \geq 1\}$ $x \in \mathbb{Z}^d$ be independent Poisson processes with rate 1, and let $\{Y_n^x, n \geq 1\}$ be independent i.i.d. sequences with $P(Y_n^x = y) = 1/2d$ for all y with $|y| = 1$. In making these definitions, we have in mind that at time T_n^x the voter at x decides for the nth time to change her mind, and she adopts the opinion of the voter at $x+Y_n^x$. Given the ingredients above, the reader could undoubtedly construct the process but, with future developments in mind, we will use a special recipe. For each x and n, we draw an arrow from $(x+Y_n^x, T_n^x)$ to (x, T_n^x), and write a δ at (x, T_x^n). (The reasons for adopting this strange notation will become clear as we proceed.) To construct the process from this "graphical representation," we imagine fluid entering the bottom at the points in ξ_0 and flowing up the structure. The δ's are dams and the arrows are pipes which allow the fluid to flow in the indicated direction. To make this definition mathematical, we say that there is a path from $(x,0)$ to (y,t) if there is a sequence of times $s_0 = 0 < s_1 < s_2 \ldots < s_n < s_{n+1} = t$ and spatial locations $x_0 = x, x_1, \ldots, x_n = y$ so that:

(i) for $i = 1,2,\ldots,n$ there is an arrow from x_{i-1} to x_i at time s_i, and

(ii) the vertical segments $\{x_i\} \times (s_i, s_{i+1})$, $i = 0,1,...,n$ do not contain any δ's.

When there is a path from $(x,0)$ to (y,t), it follows that the individual at y at time t has the same opinion as individual x at time 0. Since every individual has the same opinion as some individual at time 0, it follows that

$$\xi_t^A = \{ y : \text{for some } x \in A \text{ there is a path from } (x,0) \text{ to } (y,t)\}$$

gives the state at time t when $\xi_0^A = A$. If one wants to compute ξ_t^A from the graphical representation given above, the easiest thing to do is work backward. Suppose, for instance, we want to compute the opinion of the voter at 0 at time 1 in the realization drawn in Figure 2.1. Working backward, we see that at time t_5 she imitated the voter at -1, who in turn imitated the voter at -2 at time t_2. So the opinion at 0 at time 1 is the same as the opinion at -2 at time 0.

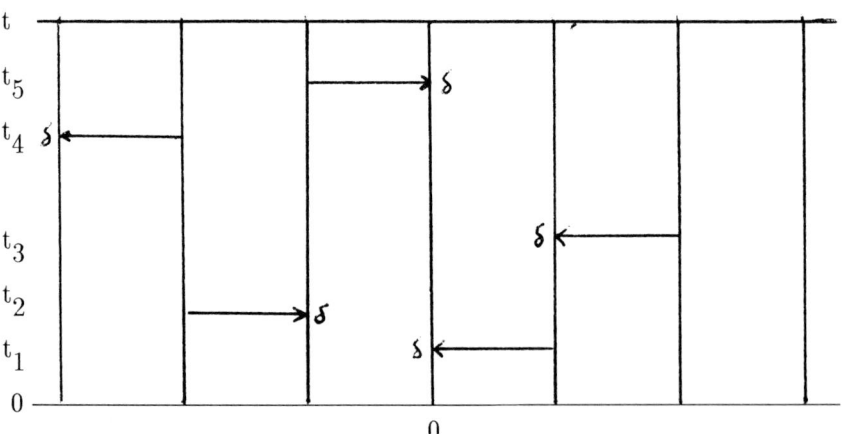

Figure 2.1

The last computation motivates the definition of a dual process for the voter model. To construct this process up to time t, reverse the direction of the arrows in the graphical representation and change time by mapping $\hat{s} = t-s$. If we define paths as before, and let

$$\hat{\xi}^B_t = \{\, x : \text{for some } y \in B \text{ there is a path from } (y,\hat{0}) \text{ to } (x,\hat{t})\},$$

then it is clear that

(1) $$\{\xi^A_t \cap B \neq \phi\} = \{\hat{\xi}^B_t \cap A \neq \phi\}.$$

The last definition is convenient for (1), but it leads to a process with very violent behavior as t varies. For instance in the realization drawn in Figure 2.1, $\hat{\xi}^{\{0\}}_t = \{1\}$ for $t_1 < t < t_5$ and $\hat{\xi}^{\{0\}}_t = \{-2\}$ for $t > t_5$.

To fix this, we define a dual process $\check{\xi}_t$ that has the same one dimensional distributions as $\hat{\xi}_t$. We begin with a graphical representation that for each $x \in \mathbb{Z}^d$, and y with $|y| = 1$, has "δ–arrows" from x to $x+y$ (i.e., a δ at x and an arrow from x to $x + y$) at rate $1/2d$. We define a "path" as before and let

$$\check{\xi}^B_t = \{\, x : \text{for some } y \in B \text{ there is a path from } (y,0) \text{ to } (x,t)\}.$$

Since the new gadgets are the same as the old ones with the arrows reversed, it should be clear that $\check{\xi}^B_t$ and $\hat{\xi}^B_t$ have the same distribution, so we have

(2) $$P(\, \xi^A_t \cap B \neq \phi \,) = P(\, \check{\xi}^B_t \cap A \neq \phi \,).$$

To see how $\tilde{\xi}_t$ behaves, we observe that the effect of a δ–arrow on the configuration is given by the table in Figure 2.2. There $\xi(x)$ indicates the state of x and 1 (resp. 0) denotes a site which is wet (resp. dry).

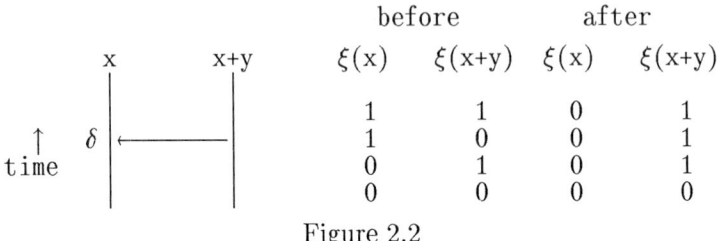

Figure 2.2

If we think of the 1's as particles and 0's as empty sites, then $\tilde{\xi}_t$ may be described as a "coalescing random walk." A δ–arrow from x to $x+y$ causes a particle at x to jump to $x+y$. If $x+y$ is occupied, the two particles coalesce to 1.

When the dual process starts from $B = \{0\}$, its behavior is trivial to compute. For any $t \geq 0$, $\tilde{\xi}_t^0$ (our abbreviation for $\tilde{\xi}_t^{\{0\}}$) has only one element and its position, X_t, is a simple random walk. X_t stays at a site for a mean one exponential amount of time, and then jumps to a randomly chosen neighbor. Combining the last observation with the duality equation (2) shows

(3) $$P(0 \in \xi_t^A) = P(X_t \in A).$$

The last formula implies, in particular, that if the initial distribution is $\nu_\theta \equiv$ product measure with density θ (each site is independently occupied with probability θ), then $P(0 \in \xi_t) = \theta$. To determine the limiting behavior of ξ_t in this case, we compute the two dimensional distributions. Let $\xi_t^A(y) = 1$ if $y \in \xi_t^A$, and $= 0$ otherwise. Now

$$P(\xi_t^A(0) \neq \xi_t^A(x)) \leq P(\zeta_t^0 \neq \zeta_t^x)$$

and the definition of the dual implies $\zeta_t^0 \cup \zeta_t^x = \zeta_t^{\{0,x\}}$, so

$$P(\xi_t^A(0) \neq \xi_t^A(x)) \leq P(|\zeta_t^{\{0,x\}}| = 2).$$

The particles that make up ζ_t^0 and ζ_t^x perform simple random walks that are independent until they hit. So if X_t and X_t' are independent simple random walks with $X_0 = 0$ and $X_0' = x$, then

$$P(|\zeta_t^{\{0,x\}}| = 2) = P(X_s \neq X_s' \text{ for all } 0 \leq s \leq t).$$

Now $\overline{X}_s = X_s' - X_s$ is itself a simple random walk which takes steps at rate 2. So if $d = 1$ or 2, the probability $\overline{X}_s$ will hit 0 is 1, and

(4) $$P(\xi_t^A(0) \neq \xi_t^A(x)) \to 0 \text{ as } t \to \infty.$$

The last result says that starting from any initial state A, the voter model approaches total consensus in $d = 1$ and 2. In $d \geq 3$, simple random walk is transient, so $P(\overline{X}_s \neq 0 \text{ for all } s) > 0$ and differences of opinion may persist. If we use ξ_t^θ to denote the voter model with initial distribution ν_θ, then

(5) $$P(\xi_t^\theta(0) \neq \xi_t^\theta(x)) = 2\theta(1-\theta)P(\zeta_t^0 \neq \zeta_t^x).$$

To see this, observe that in order for $\xi_t^\theta(0) \neq \xi_t^\theta(x)$, it is necessary that $\zeta_t^0 \neq \zeta_t^x$, and that the voters at the two sites have different opinions in ξ_0^θ, an event of probability

$2\theta(1-\theta)$. The right–hand side of (5) converges to a nonzero limit as $t \to \infty$. With a little more work, one can show that the finite dimensional distributions of ξ_t^θ converge. To do this, we begin by observing

$$P(\xi_t^\theta \cap B = \phi) = E((1-\theta)^{|\xi_t^B|}).$$

Since $|\xi_t^B|$ can only decrease, it follows that $P(\xi_t^\theta \cap B = \phi)$ increases to a limit as $t \to \infty$. The inclusion–exclusion formula can be used to write every probability of the form $P(\xi_t^\theta(x_1) = i_1,...,\xi_t^\theta(x_k) = i_k)$ where $x_1,...,x_k \in Z^d$ and $i_1,...,i_k \in \{0,1\}$, in terms of $P(\xi_t^\theta \cap B = \phi)$ with $B \subset \{x_1,...,x_k\}$. So we have shown that the finite dimensional distributions converge to those of some measure μ_θ on $\{0,1\}^{Z^d}$.

Having shown that ξ_t^θ converges to a limit μ_θ, "general nonsense" tells us that μ_θ is a stationary distribution. The details are not hard or exciting, but for completeness we will sketch a proof. First observe that on $\{0,1\}^{Z^d}$ weak convergence (denoted by $\Rightarrow$) is the same as convergence of finite dimensional distributions. Let μT_t denote the distribution of ξ_t when ξ_0 has distribution μ. A simple computation shows that if $\mu_n \Rightarrow \mu$ then $\mu_n T_t \Rightarrow \mu T_t$, and from this it follows easily that

$$\mu_\theta T_t = (\lim_{s\to\infty} \nu_\theta T_s)T_t = \lim_{s\to\infty} \nu_\theta T_{s+t} = \mu_\theta.$$

The existence of the stationary distributions μ_θ is much simpler than their uniqueness. Holley and Liggett (1975) have shown that the set of the stationary distributions is the closure (in the weak topology) of the convex hull of $\{\mu_\theta : 0 \le \theta \le 1\}$, and give conditions on the initial distribution that guarantee that $\xi_t \Rightarrow \mu_\theta$ as $t \to \infty$. A description of these results can be found in Chapter V of Liggett (1985b). Unfortunately, he chose to discuss what happens when the voter of x imitates the voter at y at rate p(x,y), where p is the transition probability of a Markov chain on a countable set S. In this level of generality, it does not make sense to define $\overline{X}_t = X_t - X'_t$, and there is a stationary distribution μ for each bounded harmonic function h. (Start with a product measure with $P(x \in \xi_0) = h(x)$, use (3), and let $t \to \infty$.) These new difficulties grab the spotlight in the proof and the simple special case we have considered here gets lost in the shuffle. A clear treatment of the simple case can be found in the original article of Holley and Liggett (1975).

Having demonstrated the dichotomy between d=2 and d=3, we will illustrate it with some computer simulations. In the first sequence of pictures we are looking at the voter model on $\{0, .. 24\}^2$ with periodic boundary conditions, starting from a product measure in which . and ∗ have equal probability. Notice the difference in texture between the configurations at times 0 and 5. As time goes on the tendency of the model to cluster becomes more pronounced. In Chapter 3 we will see that the number of ∗'s is a simple random walk run at rate (1/2) times the number of neighbors that are different, so the probability a given species emerges victorious is proportional to the fraction of sites that have that opinion.

The second set of pictures shows the state of $\{12\} \times \{0,..24\}^2$ in a voter model on $\{0,..24\}^3$ with periodic boundary conditions, starting from the same product measure. This time clustering occurs but is not as pronounced. To track the "convergence to equilibrium" we look at the fraction of neighbors that are different. If we were looking at an infinite system, this would not be random and would converge to (1/2)P(two random walks starting from adjacent points never hit). Consulting page 103 of Spitzer (1976) we see the last quantity = (1/2)(.65946267) = .3297, which agrees reasonably well with the value in the last picture.

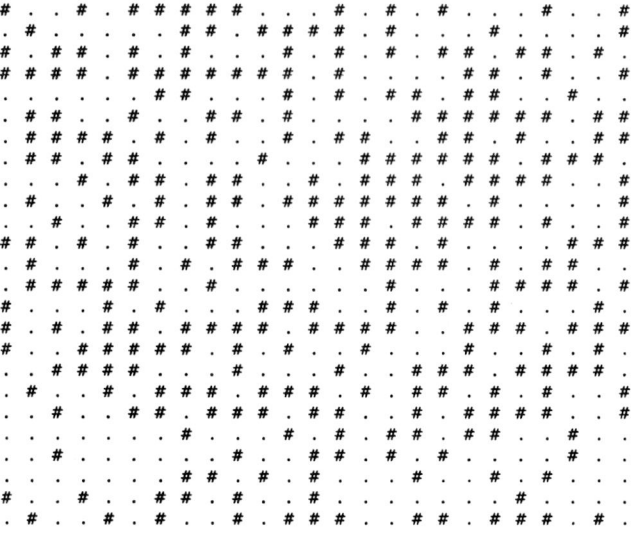

TIME = 0

. 321 # 304

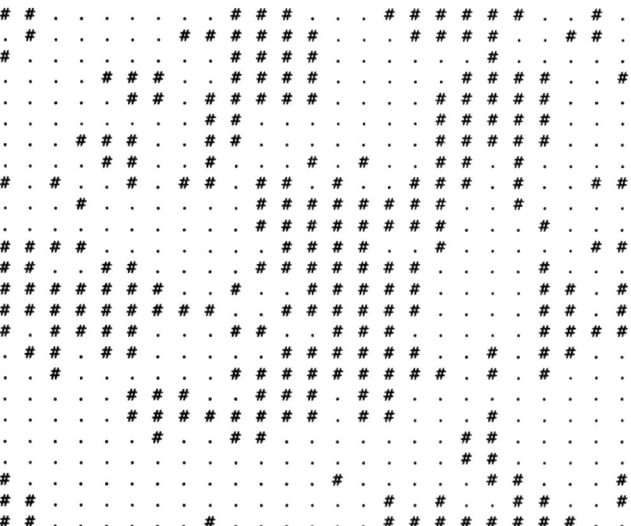

TIME = 5

. 367 # 258

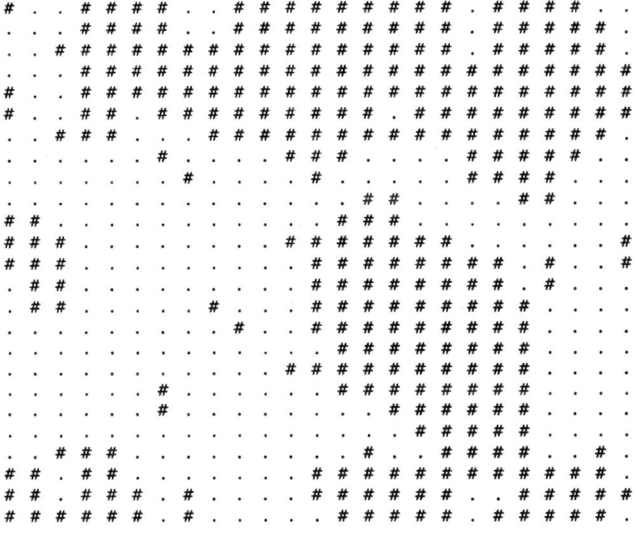

```
TIME = 25

         .  304     #   321
```

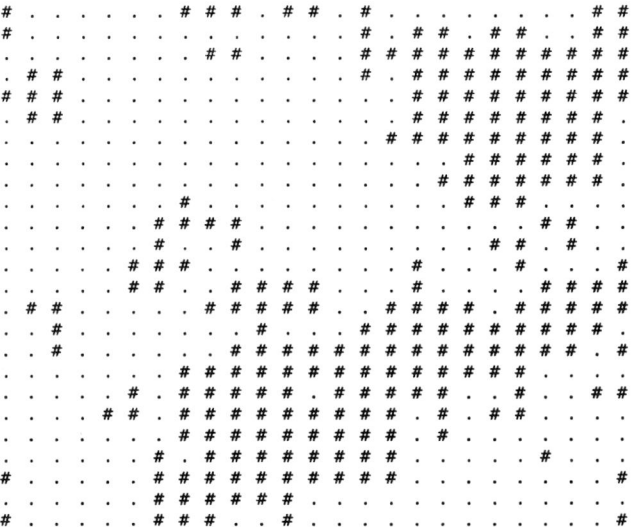

```
TIME = 125

         .  377     #   248
```

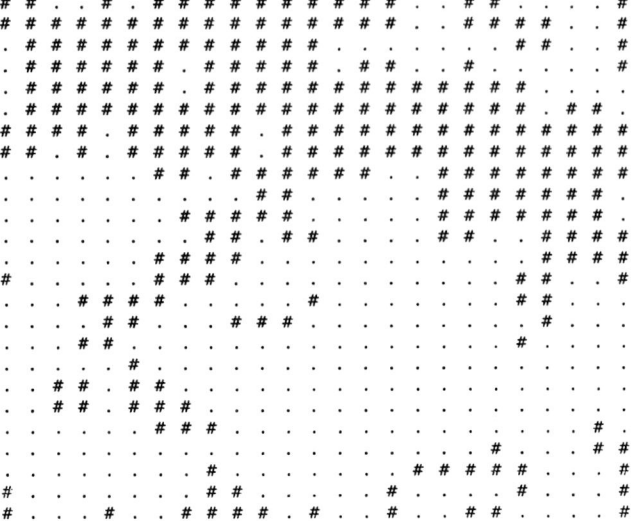

TIME = 250

. 353   # 272

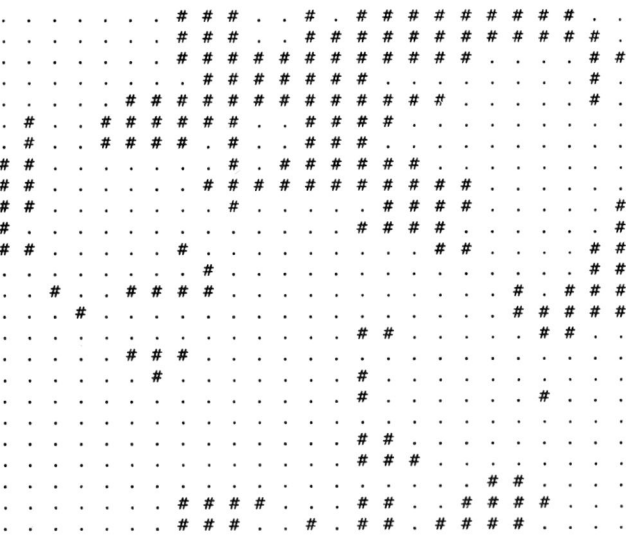

TIME = 375

. 442   # 183

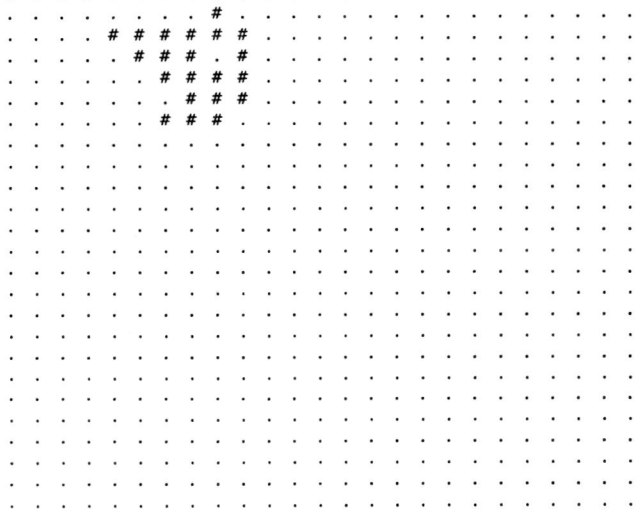

TIME = 500

. 604   # 21

TIME = 513

. 625   # 0

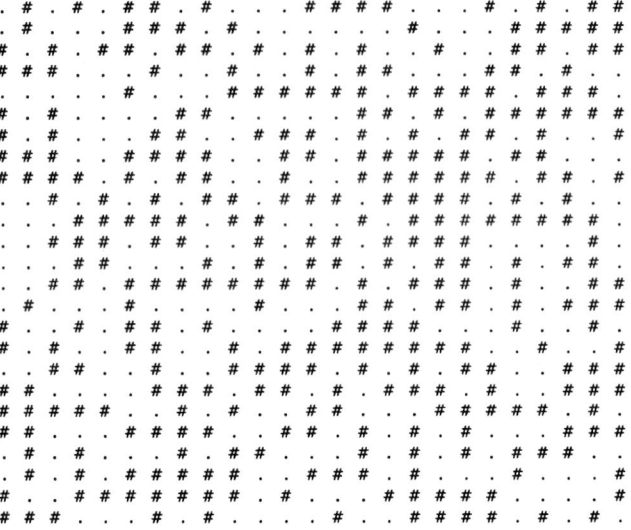

TIME = 0

P(neighbors different) = .496

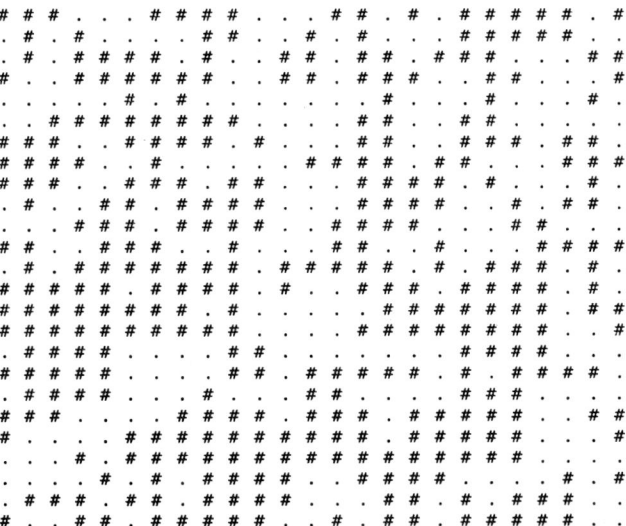

TIME = 5

P(neighbors different) = .368

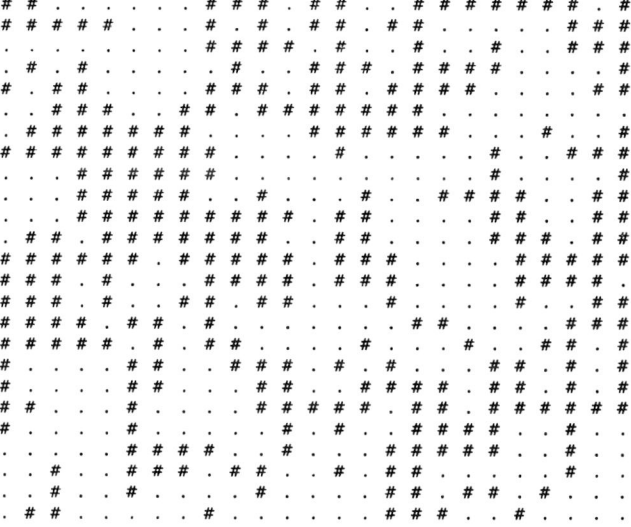

TIME = 25

P(neighbors different) = .342

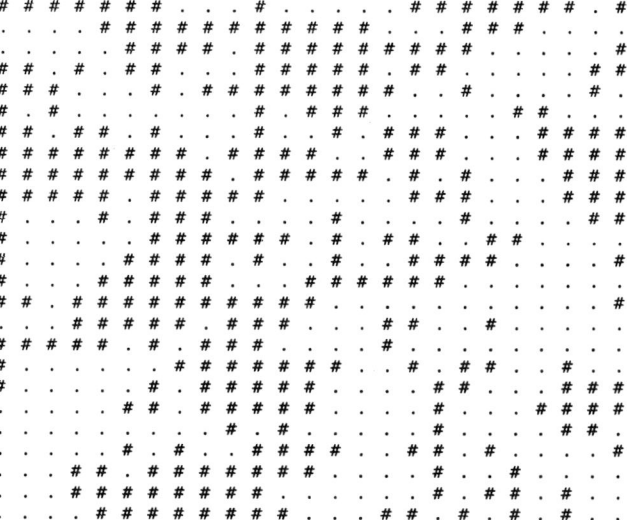

TIME = 125

P(neighbors different) = .371

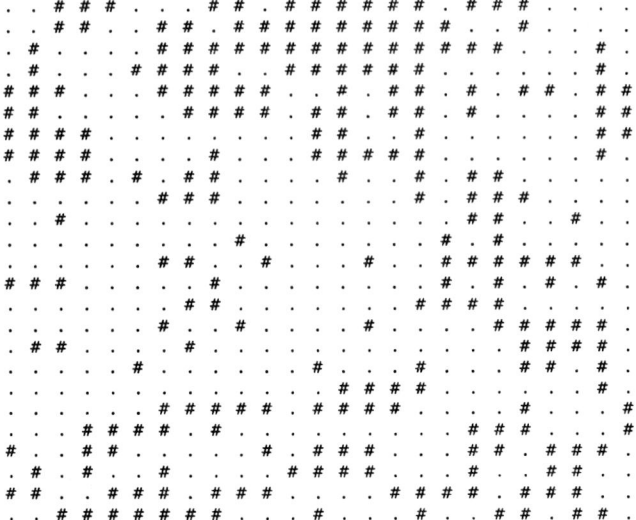

TIME = 250

P(neighbors different) = .358

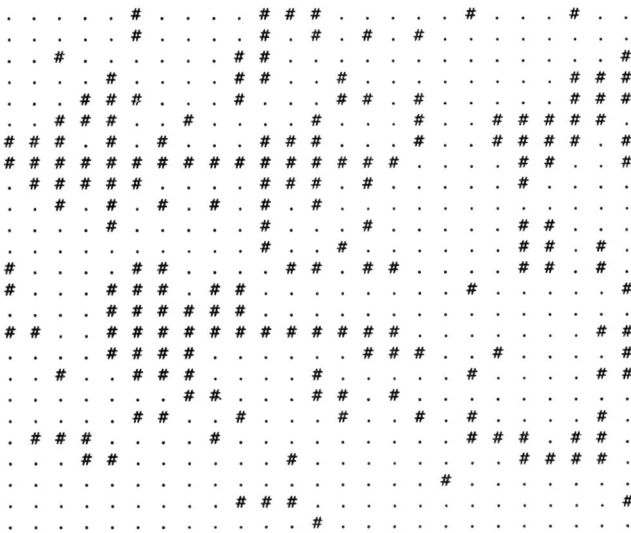

TIME = 375

P(neighbors different) = .317

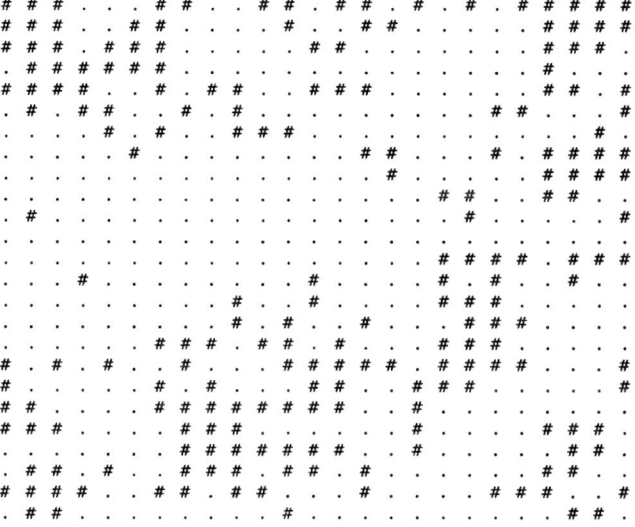

TIME = 500

P(neighbors different) = .330

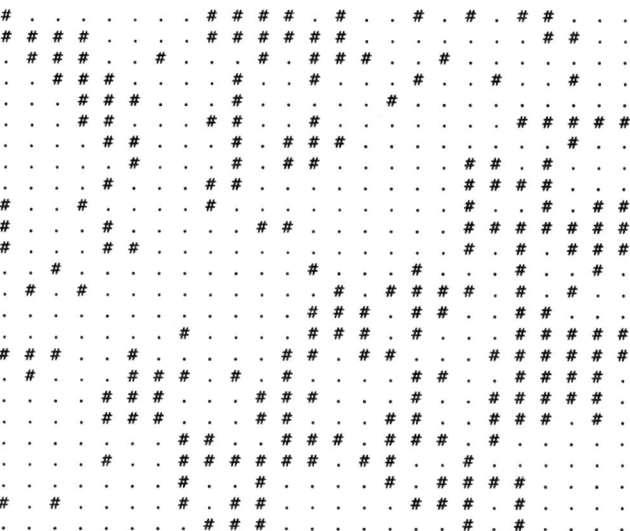

TIME = 625

P(neighbors different) = .317

3 The Biased Voter Model

In this chapter we will consider a variation of the voter model in which the people in favor of the proposition hold on to their opinions more tenaciously than the people against. Alternatively we can, as Williams and Bjerknes (1972) did, think of the points in ξ_t as being occupied by cancer cells, and the points in ξ_t^c as being occupied by healthy cells. With either interpretation in mind, we can formulate the dynamics as follows:

(i) if $x \in \xi_t$ then x becomes vacant at a rate equal to the number of vacant neighbors;

(ii) if $x \notin \xi_t$ then x becomes occupied at a rate equal to λ times the number of occupied neighbors.

If $\lambda = 1$, we get the voter model (run at 2d times the usual rate). When $\lambda > 1$, we get a model appropriate to the two situations described above. In the second interpretation, λ is called the "carcinogenic advantage." If $c < 1$ then the model with $\lambda = c$ can be obtained from the model with $\lambda = 1/c$ by relabeling occupied sites as vacant, and running the process at rate c. The last observation shows that we can take $\lambda \geq 1$ without loss of generality, and this will be convenient when we construct the process below.

Before getting into the details of the construction, the first thing to do is explain the property that makes this model easy to analyze. Let ξ_t be the process starting from a finite set and $\partial \xi_t = \{ (x,y) : |x-y| = 1, x \in \xi_t, y \notin \xi_t \}$. A little thought reveals that $|\xi_t|$ = the number of points in ξ_t, increases by 1 at rate $\lambda|\partial\xi_t|$, and decreases by 1 at rate $|\partial\xi_t|$. Observe that if $x \in \xi_t$, then x becomes vacant at rate $|\{ y : |x-y| = 1, y \notin \xi_t\}|$. If S_n is the size of ξ_t after the nth jump, then S_n is a random walk with

$$S_{n+1} = \begin{cases} S_n+1 & \text{with prob. } \lambda/(\lambda+1) \\ S_n-1 & \text{with prob. } 1/(\lambda+1). \end{cases}$$

From the last observation, we see that if ξ_t^0 is the process starting from $\xi_0^0 = \{0\}$ then

$$P(|\xi_t^0| > 0 \text{ for all } t \geq 0) = \begin{cases} (\lambda-1)/\lambda & \text{if } \lambda \geq 1 \\ 0 & \text{if } \lambda \leq 1. \end{cases}$$

So

$$\lambda_c \equiv \inf\{ \lambda : P(|\xi_t^0| > 0 \text{ for all } t \geq 0) > 0\} = 1.$$

λ_c is the "critical value" for the process, and is a notion we will see many times below, although in what follows we will very rarely know its value. The reader should note that here the critical value and survival probability are exactly the same as in a branching process in which particles give birth at rate λ and die at rate 1. That is, the Markov chain Z_t that makes transitions

$$Z_t \to \begin{cases} Z_t+1 & \text{at rate } \lambda Z_t \\ Z_t-1 & \text{at rate } Z_t. \end{cases}$$

In fact, ξ_t and Z_t are just S_n run at different rates.

The computations in the last paragraph identify λ_c. The next questions to answer are: What does the process look like when $\lambda > \lambda_c$? When $\lambda < \lambda_c$? We begin with the case $d = 1$, because in this situation, $\xi_t^0 = [\ell_t, r_t] \cap Z$ where $\ell_t = \inf \xi_t^0$ and $r_t = \sup \xi_t^0$. To see this, observe that occupied sites in the interior (ℓ_t, r_t) cannot become vacant. Looking at r_t and ℓ_t we see that when $|\xi_t^0| \geq 2$ we have

$$r_t \to \begin{cases} r_t + 1 & \text{at rate } \lambda \\ r_t - 1 & \text{at rate } 1, \end{cases} \qquad \ell_t \to \begin{cases} \ell_t - 1 & \text{at rate } \lambda \\ \ell_t + 1 & \text{at rate } 1, \end{cases}$$

and the two processes move independently until they hit. To formulate the last property, it is convenient to let R_t and L_t be two independent Markov chains starting from $R_0 = L_0 = 0$ with the transition probabilities given above, and let $\tau = \inf\{ t : R_t < L_t \}$. If we define

$$\xi_t^0 = \begin{cases} [L_t, R_t] \cap \mathbb{Z} & t < \tau \\ \phi & t \geq \tau, \end{cases}$$

then ξ_t^0 is a realization of the biased voter model starting from $\xi_t^0 = \{0\}$.

The last paragraph identifies ξ_t^0 as an interval whose endpoints perform independent random walks. As $t \to \infty$, $R_t/t \to \lambda-1$ and $L_t/t \to -(\lambda-1)$ almost surely, so we have

(1) **Theorem.** If $\lambda > \lambda_c$ then on $\Omega_\infty = \{\xi_t^0 \neq \phi \text{ for all t}\}$,

$$r_t/t \to (\lambda-1), \qquad \ell_t/t \to -(\lambda-1), \quad \text{and} \quad |\xi_t^0|/t \to 2(\lambda-1) \text{ a.s.}$$

When $\lambda < \lambda_c$, the process dies out, and interest centers on how fast this happens. Now $P(\xi_t^0 \neq \phi) \leq P(R_t \geq 0) + P(L_t \leq 0)$. (If $R_t < 0 < L_t$ then $\xi_t^0 = \phi$.)

To estimate the two terms on the right–hand side, we observe that they are equal, and use

(2) **Lemma.** Let $X_1, X_2, \ldots$ be independent and identically distributed with $EX_i < 0$ and $\varphi(\theta) = E\exp(\theta X_i) < \infty$ for some $\theta > 0$. Then there is a $\delta > 0$ so that

$$(S_n \geq 0) \leq e^{-\delta n}.$$

Proof: This is proved like (8) in Section 1. If $\theta > 0$ then

$$P(S_n \geq 0) \leq E\exp(\theta S_n) = \varphi(\theta)^n.$$

Now $\varphi(0) = 1$ and $\varphi'(0) = EX_i \in [-\infty, 0)$, so if θ is small $\varphi(\theta) < 1$.

To use (2), we let $X_k = R_k - R_{k-1}$ for $k \geq 1$. ($R_0 = 0$.) Since R_t, $0 \leq t \leq 1$ makes a Poisson number of jumps with mean $\lambda + 1$,

$$E\exp(\theta R_1) = \sum_{k=0}^{\infty} e^{-(\lambda+1)} \frac{(\lambda+1)^k}{k!} \left(\frac{e^{-\theta}}{\lambda+1} + \frac{\lambda e^{\theta}}{\lambda+1}\right)^k < \infty$$

for all θ. Since $E R_1 = \lambda - 1 < 0$, it follows that we have

(3) **Theorem.** If $\lambda < \lambda_c$, there is a $\delta > 0$ so that $P(\xi_n^0 \neq \phi) \leq e^{-\delta n}$.

(3) says "if $\lambda < \lambda_c$ then the process dies out exponentially fast". The last conclusion was proved for $d = 1$, but the proof extends easily to $d > 1$. $|\partial \xi_t| \geq 2d$

in dimension d, so for fixed $\lambda > 1$, the process dies out faster in $d > 1$. The phrase in quotation marks is one of two that the reader will hear repeated many times in these notes. The other one is "when $\lambda > \lambda_c$, ξ_t^0 grows linearly and has an asymptotic shape." (1) shows that this is true in $d = 1$. The result for $d > 1$, due to Bramson and Griffeath ((1981) and (1980b)), is

(4) **Theorem.** If $\lambda > \lambda_c$, then there is a convex set A so that on $\Omega_\infty = \{\ t : \xi_t^0 \neq \phi$ for all t} we have for any $\epsilon > 0$

$$(1-\epsilon)tA \cap \mathbb{Z}^d \subset \xi_t^0 \subset (1+\epsilon)tA \ \text{ for all t sufficiently large.}$$

It is not clear if the last result is comforting to someone with skin cancer (who is undoubtedly interested in the boring event Ω_∞^c). However, (4) gives a precise description of the asymptotic behavior of ξ_t^0, and has at least one interesting corollary.

$$\text{On } \Omega_\infty, \ |\xi_t^0|/t^d \to |A| \quad \text{a.s.,}$$

where $|A|$ = the volume of A. In proving (4), we will follow the approach of Bramson and Griffeath, using their notation as much as possible. In this section we will content ourselves to prove the main result of BG I (our shorthand for their (1983) paper).

(5) **Proposition.** Let $D(0,\delta t) = \{\ y : |y| \leq \delta t\}$. There is a $\delta > 0$ so that on Ω_∞

$$\xi_t^0 \supset D(0,\delta t) \cap \mathbb{Z}^d \ \text{ for all t sufficiently large.}$$

The reader should note that $\xi_t^0 \subset$ the process with no deaths = Richardson's model, so the radius grows at most linearly, and hence, (5) shows radius of ξ_t^0 is of

order t. The proof of the precise result (4) will have to wait until we have a little

more machinery for proving results of that type. (See Section 10d.) The proof of (5)

deserves an R rating (contains graphic details not suitable for a general audience),

and the proof of (4) is worse, so we suspect that the reader will not mind waiting. If

you get bored or confused by the proof of (5), you can skip to the end of the chapter,

and look at the pictures. The details of the proof will not be important for the

developments that follow.

We begin the proof of (5) by reformulating the result. The definitions may look

a little strange now, but the reasons for them will become clear as we proceed. Let β

> 0 be large, $\mu > 0$ be small, and for $k \geq 1$ let

$$\beta_k = 2^k\beta, \quad s_k = \beta_k/\mu, \quad t_k = s_1 + \ldots + s_k, \quad \text{and} \quad R_k = \{ \, x : \|x\|_2 < \beta_k \, \},$$

where $\|x\|_2 = (x_1^2 + \ldots + x_d^2)^{1/2}$ is the norm we will be using for the rest of the section.
We will show

(6) Lemma. If μ is chosen small enough then

$$P(\xi_t^{R(k-1)} \not\supseteq R_k \text{ for all } t \in [s_k, s_k + s_{k+1}]) \leq p_k(\beta)$$

where

$$\sum_{k=1}^{\infty} p_k(\beta) \to 0 \quad \text{as} \quad \beta \to \infty.$$

The last result implies that if ξ_t^0 contains R_0 at some time T, then with high

probability, it will contain R_k at all times $t \in [T+t_k, T+t_{k+1}]$. Since $t_{k+1} \sim$

$2\beta_{k+1}/\mu = 4\beta_k/\mu$ and R_k has radius β_k, a little arithmetic shows that (5) holds when $\delta < \mu/4$.

The proof of (6) is based on a duality that generalizes the one for the voter model. To describe the duality, we begin by constructing the biased voter model. For each x and y with $|x-y| = 1$ let $T_n^{(x,y)}$, $n \geq 1$ be a Poisson process with rate 1, and let $U_n^{(x,y)}$ be a Poisson process with rate $\lambda-1$. (Recall $\lambda > 1$.) At times $T_n^{(x,y)}$, we draw an arrow from y to x and put a δ at x. At times $U_n^{(x,y)}$, we just draw an arrow from y to x.

The gadgets used in the construction above are just a combination of the ones used for the voter model and for Richardson's model, and the process is obtained from the graphical representation as before. We define a path as we did for the voter model, and let

$$\xi_t^A = \{ y : \text{for some } x \in A \text{ there is a path from } (x,0) \text{ to } (y,t)\}.$$

Since both $\delta\!\!-\!\!$ and $\longleftarrow$ turn a 0 1 into a 1 1, but only $\delta\!\!-\!\!$ turns a 1 0 into a 0 0, and neither gadget effects a 1 1 or a 0 0, we get a process with the desired transition rates.

As in the construction of the voter model, the graphical representation gives us a bonus. If we reverse the arrows, and reverse time by mapping $\hat{s} = t-s$, we get a process

$$\hat{\xi}_t^B = \{ x : \text{for some } y \in B \text{ there is a path from } (y,\hat{0}) \text{ to } (x,\hat{t})\},$$

which has

$$\{\xi_t^A \cap B \neq \phi\} = \{\hat{\xi}_t^B \cap A \neq \phi\}.$$

As before, the sample paths of $\hat{\xi}_t^B$ have very violent behavior, so we replace $\hat{\xi}_t^B$ by a process $\check{\xi}_t^B$ constructed from a graphical representation that has δ–arrows from x to y at rate 1 if $|x{-}y| = 1$ (i.e., a δ at x and an arrow from x to y), and arrows from x to y at rate $\lambda{-}1$ (again if $|x{-}y| = 1$). Since the new gadgets are the same as the old ones with the arrows reversed,

(7) $$P(\, \xi_t^A \cap B \neq \phi \,) = P(\, \check{\xi}_t^B \cap A \neq \phi \,).$$

When $\lambda = 1$ (i.e., the voter model), the dual was a coalescing random walk. When $\lambda > 1$, particles can give birth at neighboring sites (subject to the restriction of at most one particle per site). We imagine that if a particle gives birth onto an occupied site the two particles coalesce to 1, and we call the new dual a "coalescing branching random walk." Letting $B = x$ in the duality equation (7), gives

(8) $$P(x \in \xi_t^A) = P(\check{\xi}_t^x \cap A \neq \phi),$$

and (6) can be reformulated in terms of the dual process. At first, it may look like we have traded one hard problem for another, but the dual has two nice features: $P(\check{\xi}_t^x \neq \phi$ for all t$) = 1$ and it is possible to estimate the right side of (8) by using a $\mathbb{Z}^d$ valued process $X_t^x \in \check{\xi}_t^x$. The process X_t^x can be informally described by the following rules:

(i) if X_t^x is forced to move by a δ–arrow then move it;

(ii) if the particle at X_t^x branches, then follow the branch if it takes you closer to 0.

To make this precise, we describe the transition rates for X_t^x:

$y \to z$ at rate λ if $\|z-y\|_2 = 1$ and either $\|z\|_2 < \|y\|_2$

or for some $i \le d$ $y_i = 0$ and $z_i = 1$,

$y \to z$ at rate 1 for all other z with $\|z-y\|_2 = 1$.

The unadvertised clause in the middle of the precise definition is caused by the fact that when all the coordinates of y are nonzero, half of its neighbors have smaller norms, but when, for example, $y = (2,0)$ in $d = 2$ only $(1,0)$ has a smaller norm. The definition above fixes this problem in such a way that exactly half of the neighbors satisfy the first condition, so the total rate X_t^x makes transitions is $(\lambda+1)d$. The last property is nice, but what is nicer, and more important, is that if γ is large enough $Z_t = \|X_t^x\|_2$ has negative drift when $Z_t > \gamma$. To see this, observe that a simple, but somewhat tedious, calculation shows that the drift toward 0 is minimized for sites on the axes, and the minimal drift is asymptotically $(\lambda-1)$ as $R \to \infty$. (It is for this reason that we use the norm $\| \ \|_2$. If we used our favorite norm, $|x| = |x_1| + \ldots + |x_d|$, then on the axes only one point would have smaller norm and the other $2d - 1$ would have larger norm.)

The last paragraph is the key to the proof, so it is good to take a minute to see where we stand. We want to show that

$$P(\xi_t^{R(k-1)} \supset R_k \text{ for all } t \in [s_k, s_k + s_{k+1}]) \approx 1$$

for $R_k = \{x : \|x\| < \beta_k\}$, $\beta_k = 2^k \beta$, and $s_k = 2^k \beta/\mu$. Duality tells us

(9) $\qquad P(x \notin \xi_t^{R(k-1)}) = P(\xi_t^{zx} \cap R_{k-1} = \phi) \le P(X_t^x \notin R_{k-1}).$

Intuitively, the worst case for the last quantity occurs when $t = s_k$ and $x \in \partial R_k$, since in that case the time is smallest and the distance largest. X_t^x is drifting in at rate $(\lambda-1)$, and has to go a distance $2^k \beta - 2^{k-1} \beta = \beta_k/2$ in time β_k/μ. So if $\mu < 2(\lambda-1)$ this term should be small.

To estimate $P(X_t^x \notin R_{k-1})$, we define a family of continuous time processes on $[-\gamma, \infty) \subset \mathbb{R}$ by

$$(10) \qquad\qquad Z_t^\alpha = \|X^x(t/d(\lambda+1))\|_2 - \gamma,$$

where $\alpha = \|x\|_2 - \gamma$. Each Z_t^α makes jumps at total rate 1, and if γ is large Z_t^α has a drift toward zero which is at least $((\lambda-1)/d(\lambda+1)) - \epsilon$ on $[0,\infty)$. We have changed the time scale to make our process exactly the same as the one in BG I, so we can finish things off by quoting a result from their paper.

(11) **Lemma.** Let Z_t^α be a continuous time pure jump process with mean 1 exponential holding times and $Z_0^\alpha = \alpha$. Let Y_n^α be the embedded discrete time process and $\mathscr{F}_{n-1} = \sigma(Y_0^\alpha, ..., Y_{n-1}^\alpha)$. Assume

(i) $|Y_n^\alpha - Y_{n-1}^\alpha| \le M$, and

(ii) $E(Y_n^\alpha - Y_{n-1}^\alpha | \mathscr{F}_{n-1}) \le -(\nu+\epsilon)$ on $\{ Y_{n-1}^\alpha \ge 0 \}$.

Then there are $0 < K, \delta < \infty$ which only depend on ϵ, ν, and M so that

$$P(Z_t^\alpha \ge y) \le Ke^{-\delta y}(1+e^{-\delta(\nu t - \alpha)}).$$

The proof of (11) is fairly straightforward, but it requires work, and we have nothing to add to the details given in BG I, so we will omit the proof. The form of the result should not be surprising. Because of the lower bound on the drift toward 0 (and the bound on the jumps), the stationary distribution π for Z_t^α should have $\pi(x,\infty) \le Ce^{-\delta x}$. The factor $(1+e^{-\delta(\nu t-\alpha)})$ is there to control the behavior at small times. Z_t^α starts at α and hence will require about $\alpha/(\nu+\epsilon)$ units of time to get to 0. Not coincidentally, $\exp(-\delta(\nu t-\alpha)) \le 1$ when $t \ge \alpha/\nu$.

With (11) in hand, the rest is arithmetic. Recall

$$R_k = \{\, x : \|x\| < \beta_k \,\}, \quad s_k = \beta_k/\mu, \quad \text{and} \quad \beta_k = 2^k\beta.$$

This means we want to take

$$\alpha = \beta_k - \gamma, \quad t \ge s_k = \beta_k/\mu, \quad \text{and} \quad \nu = (\lambda-1)/d(\lambda+1) - 2\epsilon,$$

which is possible if μ is chosen $< (\lambda-1)$. (Recall the time change in (10).) Picking $\beta > 2\gamma$, and setting $y = \beta_{k-1}/2$ in (11), gives

$$(12) \qquad P(x \notin \xi_t^{R(k-1)}) \le 2K\exp(-\delta\beta_{k-1}/2)$$

for $x \in R_k$ and $t \in [s_k, s_k+s_{k+1}]$.

The last result gives the bound we want for a single time. To extend the last result to all $t \in [s_k,s_k+s_{k+1}]$, we observe that if $x \notin \xi_t^{R(k-1)}$ then with probability $\ge e^{-2d\lambda}$ it will remain vacant for one unit of time, so

$$e^{-2d\lambda} P(\, x \notin \xi_t^{R(k-1)} \text{ for some } t \in [s_k,s_k+s_{k+1}]\,)$$
$$\le E|\{\, t \in [s_k,s_k+s_{k+1}+1] : x \notin \xi_t^{R(k-1)} \,\}|$$
$$\le 2K\cdot(1+s_{k+1})\cdot\exp(-\delta\beta_{k-1}/2).$$

Putting things together, we see that the probability $\xi_t^{R(k-1)}$ fails to contain $R(k)$ at some time in $[s_k, s_k + s_{k+1}]$ is less than

(13) $$|R_k| \cdot e^{2d\lambda} \cdot 2K \cdot (1 + s_{k+1}) \cdot \exp(-\delta\beta_{k-1}/2) \leq Ce^{-\gamma\beta(k)},$$

where $C, \gamma \in (0, \infty)$ are independent of β. (Recall K and δ are the constants of (11), $R_k = \{ x : \|x\| < \beta_k \}$, $s_k = \beta_k/\mu$, and $\beta_k = 2^k\beta$.) We are now done with the proof of (5).

Having waded through a morass of details, it is satisfying to note that the constants we have come up with are not too bad. In the derivation of (12), we picked $\mu < (\lambda-1)$, which leads to $\delta < (\lambda-1)/4$ in (5). In d=1, $A = (-(\lambda-1), \lambda-1)$ and the model grows faster in $d > 1$, so we leave it to the reader to figure out how and where we lost factor(s) of 2. A more interesting problem is to find upper bounds on $\alpha_d(\lambda) =$ the radius of A, when the parameter is λ in dimension d. A trivial upper bound comes from comparing with Richardson's model, but what we have in mind is something that tells how fast $\alpha_d(\lambda) \to 0$ as $\lambda \to 1$. To be specific, is $\alpha_d(\lambda) \leq C(\lambda-1)$ or is it $\approx C(\lambda-1)^\sigma$ where $\sigma < 1$? To indicate that there may be something interesting to prove here, we would like to mention a result of Biggins (1978). He proved an asymptotic shape theorem for branching random walk. If we consider the special case where particles die at rate 1 and give birth at a randomly chosen neighbor at rate λ, then the radius of the limiting set $\approx (\lambda-1)^{1/2}$ as $\lambda \downarrow 1$. This shows $\sigma \geq 1/2$ but also indicates that σ might be < 1 (the value in one dimension). Looking in my crystal ball I see $\sigma = 1/2$ in $d \geq 3$ and logarithmic corrections in $d = 2$. Like Fermat, however, my proof won't fit in the margin.

As usual, we end the section with some computer simulations. The pictures show the biased voter model with $\lambda = 2$, viewed at the first time it reaches the boundary of the box $\{ x : \| x \|_\infty \leq R \}$ of radius $R = 20, 40$, and 80. Notice that, as the theorem predicts, the times roughly double from one picture to the next.

TIME 22

TIME 45

TIME 92

4 The Contact Process

4a Edge Speeds Characterize λ_c

The contact process is a crude model of the spread of a disease or a biological population. As in the first three models we have considered, the state of the process $\xi_t \subset Z^d$, and the points in ξ_t are thought of as being occupied; or when we are thinking about the spread of a disease, they are called infected. The birth rate is the same one we have used before. The death rate is different but much simpler.

(i) If $x \notin \xi_t$, then x becomes occupied at a rate equal to λ times the number of occupied neighbors.

(ii) If $x \in \xi_t$, then x becomes vacant at rate 1.

Our first step is to construct the process. For each x and y with $|x-y| = 1$, let $\{T_n^{(x,y)}, n \geq 1\}$ be a Poisson process with rate λ, and let $\{U_n^x, n \geq 1\}$ be a Poisson process with rate 1. At times $T_n^{(x,y)}$, we draw an arrow from x to y to indicate that if x is occupied then y will become occupied (if it is not already). At times U_n^x, we put a δ at x. The effect of a δ is to kill the particle at x (if one is present). This is the fourth and last time we will construct a process from a graphical representation. The reader should note that each time the gadgets are δ's, arrows, or a combination of the two.

To construct the contact process from this graphical representation, we use the recipe we used for the voter model. We repeat the definitions from Section 2 here for the convenience of the reader. We say there is a path from $(x,0)$ to (y,t) if there is a sequence of times $s_0 = 0 < s_1 < s_2 \ldots < s_n < s_{n+1} = t$ and spatial locations $x_0 = x$, $x_1, \ldots, x_n = y$ so that:

(i) for $i = 1,2,\ldots,n$ there is an arrow from x_{i-1} to x_i at time s_i;

(ii) the vertical segments $\{x_i\} \times (s_i, s_{i+1})$, $i = 0, 1, \ldots, n$ do not contain any δ's.

To define the process starting from initial configuration A, we let

$$\xi_t^A = \{ \, y : \text{for some } x \in A \text{ there is a path from } (x,0) \text{ to } (y,t)\}.$$

Since the arrows in a path indicate births, and the absence of δ's indicate that the particles involved did not die before they gave birth, it is easy to see that the recipe above gives the contact process.

The first, and most basic, question to answer about the contact process is "When is $P(|\xi_t^0| > 0$ for all t$) > 0$?" or, in words, "For what values of the birth rate does the species have positive probability of not becoming extinct?" It is easy to see that if $\lambda \leq 1/2d$, then the process will die out. In this case, when $|\xi_t| = k$ the population decreases by 1 at rate k, and increases by 1 at rate $\leq 2d\lambda k$, the maximum birth rate occurring when no two particles in ξ_t are adjacent. It is much harder to establish that the other alternative can occur; that is, if λ is large then $P(|\xi_t^0| > 0$ for all t$) > 0$.

The difficulty in proving that the process can survive is that when $|\xi_t| = k$ the death rate is always k, but in one dimension the birth rate may be as small as 2λ if ξ_t is an interval. Of course, it is incredibly pessimistic to think that ξ_t is an interval but.... We invite the reader to think about how to prove that the contact process survives for large λ before we give one solution in Section 5a. Attacking this problem is a good way to learn first hand that "in this subject it is hard to do explicit computations." To encourage you to try to find your own solution, we would like to say that there are at least four different ways to prove the result, but finding a new one might be a valuable contribution.

For the moment then, we will ask the reader to believe that if we let

$$\lambda_c = \inf\{ \, \lambda : P(|\xi_t^0| > 0 \text{ for all t}) > 0 \, \},$$

then $\lambda_c < \infty$, and we will turn our attention to the topic that is the title to the section — finding a way to characterize λ_c. The key to the study of the contact process in one dimension is the right edge $r_t = \sup \xi_t^{(-\infty,0]}$, where $\xi_t^{(-\infty,0]}$ indicates the contact process with initial state $\{0,-1,-2,...\}$. If we let

$$r_{s,t} = \sup\{ \, y-r_s : \text{there is a path from (x,s) to (y,t) for some } x \leq r_s \}$$

then $r_{0,t} = r_t$ (since $r_0 = 0$), but what is much more important

(1) $$r_{0,s} + r_{s,t} \geq r_{0,t}$$

so $r_{s,t}$ is "subadditive."

To see the last inequality, observe that $r_{0,s} + r_{s,t}$ is the position of the right edge at time t if we pretend that all the sites $x \leq r_s$ are occupied at time s. From the last observation, and the fact that the system is translation invariant, it should be clear that $r_{s,t}$ is independent of $r_{0,s}$, and has the same distribution as $r_{0,t-s}$. So

$$\{r_{(n-1)k,nk}, \, n \geq 1\} \text{ is an ergodic stationary sequence for each } k.$$

(The variables are i.i.d.) It is even easier to see that

$$\{r_{m,m+k}, \, k \geq 0\} \overset{d}{=} \{r_{m+1,m+k+1}, \, k \geq 0\},$$

and $Er_{0,1}^+ < \infty$. For the last claim, observe that $r_t \leq R_t$, the position of the right edge in a process with no deaths, and R_t has a Poisson distribution with mean λt. With the conditions above verified, we can apply the subadditive ergodic theorem ((5) in Section 1) to conclude that

(2) **Theorem.** $\lim_{t \to \infty} r_t/t = a(\lambda)$ almost surely, where $a(\lambda) = \inf_{t>0} Er_t/t.$

The main reason for interest in the "edge speed," $a(\lambda)$, is that it enables us to characterize λ_c.

(3) **Theorem.** $\lambda_c = \inf\{\lambda : a(\lambda) > 0\} = \sup\{\lambda : a(\lambda) < 0\}.$

The first step in proving this is to show:

(4) **Lemma.** If $a(\lambda) < 0$, then $\lambda < \lambda_c$.

Proof: From the construction of the contact process, $\xi_t^0 \subset \xi_t^{(-\infty,0]}$. So if we let $r_t^0 = \sup \xi_t^0$, then $r_t^0 \leq r_t$. Let $\ell_t = \inf \xi_t^{[0,\infty)}$ and $\ell_t^0 = \inf \xi_t^0$. Repeating the reasoning above, we see $\ell_t^0 \geq \ell_t$. But if $a(\lambda) < 0$, then with probability 1,

$$\lim_{t \to \infty} r_t/t = a(\lambda) < -a(\lambda) = \lim_{t \to \infty} \ell_t/t,$$

and it follows that for t large

$$\sup \xi_t^0 = r_t^0 \le r_t < \ell_t \le \ell_t^0 = \inf \xi_t^0.$$

The last equality is only possible if $\xi_t^0 = \phi$.

(4) tells us that if the edges drift inward, then the process dies out. To prove that outward drift gives a positive probability of survival, we need the following "coupling result" which is a special feature of the nearest neighbor model.

(5) Lemma. Suppose $0 \in A$. On $\{\xi_t^0 \ne \phi\}$, $\xi_t^0 = \xi_t^A \cap [\ell_t^0, r_t^0]$.

Proof: We check that every transition preserves this equality.   If we use 1's to designate occupied sites, and 0's vacant sites, then the situation at time t might be

$$
\begin{array}{c}
\xi_t^A \\
\xi_t^0
\end{array}
\quad
\begin{array}{ccc|ccccccc|ccc}
0 & 1 & 1 & 1 & 0 & 1 & 1 & 0 & 0 & 1 & 0 & 1 & 1 \\
0 & 0 & 0 & 1 & 0 & 1 & 1 & 0 & 0 & 1 & 0 & 0 & 0.
\end{array}
$$
$$\qquad\qquad \ell_t^0 \qquad\qquad\qquad r_t^0$$

If a change occurs at a site x with $\ell_t^0 < x < r_t^0$, then the change is the same in both processes, since the state of x−1, x, and x+1 is the same in both processes. If the 1 at ℓ_t^0, or r_t^0 dies because of a δ, this happens in both processes. Finally, if there is an arrow from ℓ_t^0 to ℓ_t^0-1 or from r_t^0 to r_t^0+1, then the edge moves over by 1 and the new site is occupied in both processes. The argument above breaks down if the birth rate at x depends upon the state of x−2 and x+2. In this case $\ell_t^0 + 1$ might become occupied in the top process while it is vacant in the bottom one.

Extending the reasoning in the proof of (5), it is easy to show that on $\{\xi_t^0 \ne \phi\}$

(6a)
$$\xi_t^0 = \xi_t^{(-\infty,0]} \cap [\ell_t^0, \infty).$$

(6b)
$$\xi_t^0 = \xi_t^{[0,\infty)} \cap (-\infty, r_t^0].$$

Here one picture is worth a hundred words.

$$
\begin{array}{llll|llllllllll}
\xi_t^{(-\infty,0]} & 0 & 1 & 0 & 1 & 0 & 1 & 0 & 1 & 1 & 0 & 0 & 0 \\
\xi_t^0 & 0 & 0 & 0 & 1 & 0 & 1 & 0 & 1 & 1 & 0 & 0 & 0 \\
& & & & \ell_t^0
\end{array}
$$

As in the proof of (5), one checks that every transition preserves the equalities. This time the proof is left to the reader.

From the observations above, it follows that

(6c)
$$\ell_t^0 = \ell_t \quad \text{and} \quad r_t^0 = r_t \quad \text{on} \quad \{\xi_t^0 \neq \phi\},$$

and if we let $\tau^0 = \inf\{ \ t : \xi_t^0 = \phi\}$, then

(6d)
$$\tau^0 = \inf\{ \ t : r_t^0 < \ell_t^0\} = \inf\{ \ t : r_t < \ell_t\}.$$

Having identified τ^0 as the first time $\ell_t > r_t$, it is easy to prove that $\alpha(\lambda) > 0$ implies $P(\tau^0 = \infty) > 0$. $\alpha(\lambda) > 0$ implies $r_t \to \infty$ almost surely, and hence, there is an integer M so that

$$P(r_t \geq -M \text{ for all } t) \geq .51, \quad \text{and} \quad P(\ell_t \leq M \text{ for all } t) \geq .51,$$

the second conclusion following from symmetry. A simple argument imitating (5) and (6) shows that if we let $\ell_t^{[-M,M]} = \inf \xi_t^{[-M,M]}$ and $r_t^{[-M,M]} = \sup \xi_t^{[-M,M]}$ then

$$P(\xi_t^{[-M,M]} \neq \phi \text{ for all } t) \geq P(\ell_t^{[-M,M]} \leq 0 \leq r_t^{[-M,M]} \text{ for all } t)$$
$$= P(\ell_t^{[-M,\infty)} \leq 0 \leq r_t^{(-\infty,M]} \text{ for all } t) \geq .02,$$

and it follows that $P(\xi_t^0 \neq \phi \text{ for all } t) \geq .02/(2M+1)$. (In order for $\xi_t^{[-M,M]}$ to survive, some ξ_t^x with $x \in [-M,M]$ must survive.)

At this point, we have shown $\sup\{\lambda : \alpha(\lambda) < 0\} \leq \lambda_c \leq \inf\{\lambda : \alpha(\lambda) > 0\}$. To complete the proof, we have to rule out the possibility that $\{ \lambda : \alpha(\lambda) = 0 \}$ is an interval of positive length. The first step in doing this is to show

(7) **Lemma.** Let $r_t^A = \sup \xi_t^A$. If A and B are infinite sets with $B \subset A \subset (-\infty,-1]$, then for any finite set C

$$0 \leq r_t^{A \cup C} - r_t^A \leq r_t^{B \cup C} - r_t^B.$$

In words, the smaller process is helped more by the extra particles.

Proof: The construction of the process implies $\xi_t^{A \cup C} = \xi_t^A \cup \xi_t^C$ and $\xi_t^{B \cup C} = \xi_t^B \cup \xi_t^C$ (this property is called additivity), so

$$r_t^{A \cup C} - r_t^A = (r_t^C - r_t^A)^+,$$

and

$$r_t^{B \cup C} - r_t^B = (r_t^C - r_t^B)^+.$$

Now $r_t^A \geq r_t^B$, and $z \to (r_t^C - z)^+$ is decreasing, so the result follows. (The assumptions

on A, B, and C are needed so that $-\infty < r_t^A, r_t^B, r_t^{AUC}, r_t^{BUC} < \infty$ and the differences

are well defined.)

Letting $A = (-\infty, -1]$ and $C = \{0\}$ in (7), gives that for all infinite B $\subset$

$(-\infty, -1]$

(8)
$$E(r_t^{BU\{0\}} - r_t^B) \geq E(r_t^{(-\infty, 0]} - r_t^{(-\infty, -1]}) = 1,$$

since the system is translation invariant. The last inequality says that if we add a 1 to

the right of all the ones in the initial configuration, then this increases the expected

location of the rightmost 1 by at least 1. Using this result it is easy to show

(9) Lemma. Let $\alpha_t(\lambda) = Er_t$ when the birth rate is λ. Then

$$\alpha_t(\lambda + \delta) - \alpha_t(\lambda) \geq \delta t.$$

Proof: Construct copies of the process with parameters $\lambda + \delta$ and λ starting from

$(-\infty, 0]$ on the same space, by using Poisson processes $\{T_n^{(x,y)} : n \geq 1\}$ with rate $\lambda + \delta$,

and then flipping independent coins with probability $\lambda / (\lambda + \delta)$ of heads to determine if

the arrow will occur in the process with parameter λ. (Recall that if we thin a Poisson

process with rate μ by flipping a coin with probability p of heads to determine which

points to keep, then the result is a Poisson process with rate μp.) Let $\xi_t^{\lambda + \delta}$ and ξ_t^λ

denote these processes, let $r_t^{\lambda + \delta} = \sup \xi_t^{\lambda + \delta}$, and let $r_t^\lambda = \sup \xi_t^\lambda$. Let $\sigma = \inf\{s \geq 0 :$

$r_s^{\lambda + \delta} > r_s^\lambda\}$ be the first time the right edge of the process with parameter $\lambda + \delta$ gets

ahead. To use (8) to compare $r_t^{\lambda + \delta}$ and r_t^λ on $\{\sigma \leq t\}$, it is convenient to introduce

a third process $\bar{\xi}_s$ which $= \xi_s^{\lambda + \delta}$ for $s \leq \sigma$, and then evolves like the contact process with parameter λ. If we let $\bar{r}_t = \sup \bar{\xi}_t$ then

$$E(r_t^{\lambda+\delta}-r_t^{\lambda}) \geq E(\bar{r}_t-r_t^{\lambda};\sigma \leq t).$$

Now at time σ, $\bar{r}_\sigma = r_\sigma^{\lambda+\delta} \geq r_\sigma^{\lambda}+1$, and $\bar{\xi}_\sigma \supset \xi_\sigma^{\lambda}$, so applying (8) gives

$$E(\bar{r}_t-r_t^{\lambda};\sigma \leq t) \geq P(\sigma \leq t).$$

Since the birth rate at $r_t^{\lambda+\delta}+1$ in $\xi_t^{\lambda+\delta}$ is always $\lambda+\delta$, while the birth rate at $r_t^{\lambda}+1$ in ξ_t^{λ} is λ, it follows that

$$P(\sigma \leq t) \geq 1-e^{-\delta t}.$$

($r_t^{\lambda+\delta}$ can also get ahead when the right edges fall back.) Combining the last three displays, gives

$$E(r_t^{\lambda+\delta}-r_t^{\lambda}) \geq 1-e^{-\delta t}.$$

To strengthen the last result to the desired conclusion, write

$$\alpha_t(\lambda+\delta) - \alpha_t(\lambda) = \sum_{k=1}^{n} \alpha_t(\lambda+\delta k/n) - \alpha_t(\lambda+\delta(k-1)/n)$$

$$\geq n(1-e^{-\delta t/n}) \to \delta t$$

as $n \to \infty$. This completes the proof of (9), and of the characterization (3).

Having established the result that will be the key to the study of the one dimensional contact process, we will now relax for a minute and look at some computer simulations. The pictures show the contact process with λ = 1.35, 1.5, 1.65, 1.8, and 1.95 run for $0 \leq t \leq 100$. The simulations have been constructed so that ξ_t^λ is an increasing in λ. If numerical results, described in the next paragraph, are to be believed the third picture shows the contact process near the critical value. This seems to be confirmed by the fourth and fifth pictures, which show the edges drifting outward.

For numerical results concerning λ_c see Brower, Furman, and Moshe (1978), and Grassberger and de la Torre (1979). The first paper studies the Reggeon Quantum Spin Model, a process that the second paper explains is equivalent to the contact process. The RQSM is parameterized by a temperature $T = 1/\lambda$. On page 216 of BFM(1978) the value T_c = .60628 ± .00004 appears. Inverting gives λ_c = 1.6494. They also give the critical value for the two dimensional model: T_c = 2.428 ± .003 or λ_c =.412.

There are also rigorous results concerning λ_c. Since the right edge moves ahead one at rate λ and falls back at least one at rate 1, $\alpha(\lambda) \leq \lambda-1$. Combining this observation with (3) shows that $\lambda_c \geq 1$. This is the first of a series of bounds that can be obtained by approximating the right edge by a process in which the sites $x < r_t-k$ are assumed to be always occupied. The bound above corresponds to k=0. When k=1 a little work shows $\lambda_c \geq 1.18$. When k=2, with a little patience, one finds $\lambda_c \geq 1.28$. When k=14, the computer says $\lambda_c \geq 1.539$. See Ziezold and Grillenberger (1985) or page 289 of Liggett (1985b).

The story on upper bounds is short and sweet. Holley and Liggett (1978) have shown that $\lambda_c \leq 2/d$ in dimension d. See Section VI.1 of Liggett (1985b) for an improved version of their proof.

$$\lambda = 1.35$$

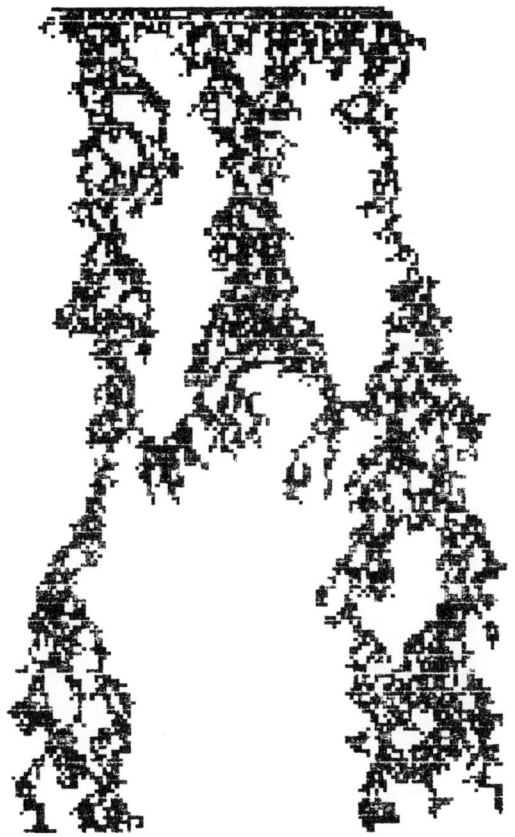

$$\lambda = 1.50$$

$$\lambda = 1.65$$

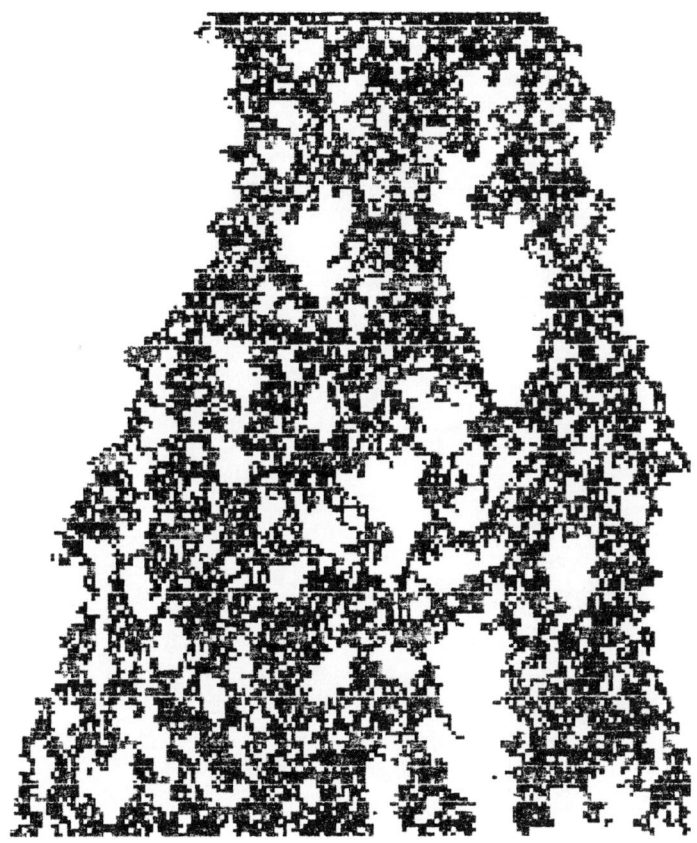

$\lambda = 1.80$

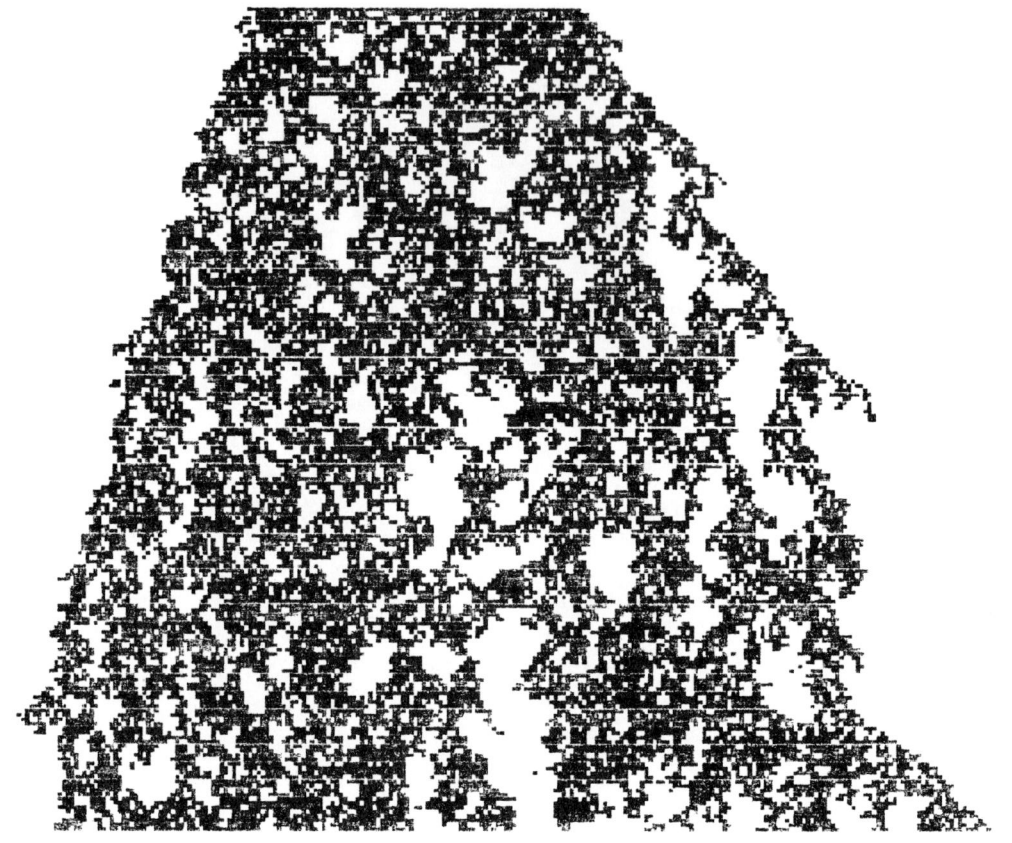

$$\lambda = 1.95$$

4b First Results in d = 1

In this section we apply the characterization developed in the last section to study the contact process in one dimension. The word first in the title of this section refers to the fact that the results here are due to Durrett (1980) and Griffeath (1981). In Chapter 10 we develop the more refined results that came later.

We begin with $\lambda < \lambda_c$. In this case the main thing to be shown is that "the process dies out exponentially fast."

(1) Theorem. If $\lambda < \lambda_c$ then there is a constant $\gamma > 0$ (that depends on λ) so that

$$P(\,\xi_t^0 \neq \phi) \leq e^{-\gamma t},$$

and as $t \to \infty$

$$\frac{1}{t} \log P(\,\xi_t \neq \phi\,) \to -\gamma.$$

Proof: When $\xi_t^0 \neq \phi$ we have $|\xi_t^0| \geq 1$ so

$$P(\,\xi_{t+s}^0 \neq \phi \mid \xi_t^0 \neq \phi\,) \geq P(\,\xi_s^0 \neq \phi\,),$$

and hence

$$P(\,\xi_{t+s}^0 \neq \phi\,) \geq P(\,\xi_t^0 \neq \phi\,)P(\,\xi_s^0 \neq \phi\,).$$

(In the language of reliability the contact process ξ_t^0 is "used better than new.") If we let $a_t = \log P(\,\xi_t^0 \neq \phi\,)$, then the last inequality implies

$$a_{t+s} \geq a_t + a_s,$$

(i.e., a_t is "superadditive"). Turning the proof of (3) in Chapter 1 upside down, shows that as $t \to \infty$

$$a_t/t \to \sup_{s>0} a_s/s \leq 0.$$

If we let $-\gamma = \sup a_s/s$, then we have proved everything except that $\gamma > 0$. To do this, we have to use the fact that $\lambda < \lambda_c$. Everything thus far is valid for all λ. If $\lambda < \lambda_c$ then $a(\lambda) = \inf Er_t/t < 0$, so there is a T with $Er_T < 0$. From subadditivity it follows that

$$r_{0,mT} \leq r_{0,T} + r_{T,2T} + \cdots + r_{(m-1)T,mT}.$$

The right–hand side, which we will call S_m, is a random walk with $ES_1 < 0$. By comparing with a process with no deaths, in which the position of the right edge R_t has a Poisson distribution with mean λt, we see that $E(\exp(\theta S_1)) \leq E(\exp(\theta R_T))$ for all $\theta > 0$. So it follows from (2) in Chapter 3 that there is a $\delta > 0$ so that

$$P(\, S_m \geq 0\,) \leq e^{-\delta m}$$

To prove that a similar bound holds for $P(\, \xi_t^0 \neq \phi\,)$, we observe that if $r_t < 0 < \ell_t$, then $\xi_t^0 = \phi$ (see (4) in Section 4a for more details), so

$$P(\, \xi_t^0 \neq \phi\,) \leq P(\, \ell_t \leq 0\,) + P(\, r_t \geq 0\,) = 2\, P(\, r_t \geq 0\,).$$

Combining this with the last two displays completes the proof.

Remark: For developments in Section 10 we would like the reader to observe that a simple modification of the argument above proves:

(2) Lemma. If $a > \alpha(\lambda)$ there are constants $C, \gamma \in (0,\infty)$ so that $P(r_t \geq at) \leq Ce^{-\gamma t}$.

Proof: Pick T so that $Er_T < a$ ...

Using (1), it is easy to show:

(3) Theorem. If $\lambda < \lambda_c$ then as $t \to \infty$ $\frac{1}{t} \log(-r_t) \to \gamma$, where γ is the constant in (1).

The reader should note that this says $r_t \to -\infty$ exponentially fast, so $\alpha(\lambda) = -\infty$ for $\lambda < \lambda_c$. This result is not important for what follows and the reader can safely skip the proof.

Proof: Let $c_n = 1/P(\xi_n^0 \neq \phi)$. We want to show $r_n \approx -c_n$. The upper bound is easy. Let $b_n = c_n/n^2$, and let $U_n = \sup \{ x \leq 0 : \xi_n^x \neq \phi \}$.

$$P(U_n > -b_n) \leq b_n P(\xi_n^0 \neq \phi) = 1/n^2$$

so the Borel–Cantelli lemma implies

$$P(U_n > -b_n \text{ i.o. }) = 0,$$

where, as usual, i.o. stands "for infinitely often." To estimate the location of r_n, we consider $s_n = \sup \xi_n^{(-\infty, b(n)]}$ and compare with a contact process with no deaths. If R_n is a Poisson random variable with mean $n\lambda$ then applying (2) in Chapter 3 to $R_n - 2\lambda n$ gives

$$P(\, R_n > 2\lambda n) \le e^{-\delta n},$$

where $\delta > 0$. Combining this with the previous displayed equation gives

$$P(\, r_n > -b_n + 2\lambda n \text{ i.o. }) = 0.$$

Since $b_n = 1/n^2 P(\, \xi_n \ne \phi)$, the last result implies

$$\liminf_{n \to \infty} \log(-r_n)/n \ge \gamma.$$

To prove the opposite inequality, we need to find enough independent events. The two keys to doing this are in the first part of the proof; the contact process spreads more slowly than a rate λ Poisson process, and if we can show $r_n > -Anc_n$ where $A < \infty$ is a constant, that will be good enough. We begin by observing that if we let $\varphi(\theta) = E\exp(\theta R_1)$, where R_n is the right edge of the process with no deaths introduced above, then

$$P(\, R_n \ge Kn\,) \le e^{-Kn} \varphi(1)^n \le e^{-2\gamma n}$$

if K is large. Let L be the smallest integer $\ge Kn$. (Observe that L depends on n but the notation used does not reflect this.) Consider a modification of the contact process that starts with { 0, −2L, −4L, ... } occupied, and in which births at L, −L, −3L, −5L, ... are not allowed. (This makes the processes in the different compartments indpendent.) We will call a point of the form x = −2mL good if $\xi_n^x \neq \phi$, and the process has not tried to give birth onto x+L or x−L. If n is large, the probability x is good is at least

$$P(\, \xi_n^0 \neq \phi) - 2P(\, R_n \geq Kn \,) \geq P(\, \xi_n^0 \neq \phi \,)/2.$$

(Recall L ≥ Kn.) Let $V_n = \inf\{\, m \geq 0 : x = -2mL$ is good $\}$. The events that appear in the definition of V_n are independent (for the modified contact process introduced above), so

$$P(\, V_n < -nc_n \,) \leq (1 - 1/2c_n)^{nc(n)}.$$

(Recall $c_n = 1/P(\, \xi_n^0 \neq \phi)$.) As n → ∞, $(1 - 1/2c_n)^{c(n)} \to e^{-1/2}$, so if n is large

$$P(\, V_n < -nc_n) \leq e^{-n/3}.$$

Since $r_n \geq -2LV_n - L$, it follows from the Borel–Cantelli lemma that

$$P(\, r_n < -2(Kn + 1)(nc_n+1) \text{ i.o. }) = 0.$$

The last result implies

$$\limsup_{n \to \infty} \log(-r_n)/n \leq \gamma,$$

and completes the proof. Readers who are sticklers for detail will have noticed that we proved the result for n → ∞ (through the integers), but have claimed the result for t → ∞.

Given the developments above, the details necessary to handle intermediate times are not difficult and are left to the reader.

With (3) established, we turn now to study the behavior of the process for $\lambda > \lambda_c$. The first step is to introduce a dual process similar to the ones used in Chapters 2 and 3. This part works in any dimension. We reverse the arrows, and reverse time by mapping $\hat{s} = t-s$, to get a process

$$\hat{\xi}^B_s = \{x : \text{for some } y \in B \text{ there is a path from } (y, \hat{0}) \text{ to } (x, \hat{s})\}$$

that has

$$\{ \xi^A_t \cap B \neq \phi \} = \{ \hat{\xi}^B_t \cap A \neq \phi \}.$$

As in the two previous cases, we replace $\hat{\xi}^B_t$ by a process $\check{\xi}^B_t$ constructed from a graphical representation that has arrows from x to y at rate λ if $|x-y| = 1$, and δ's at rate 1. Since the new gadgets are the just the old ones with the directions of the arrows reversed, $\hat{\xi}^B_t$ and $\check{\xi}^B_t$ have the same distribution. Since the rate of arrows from x to y is the same as the rate of arrows from y to x, $\check{\xi}^B_t$ and ξ^B_t have the same distribution. Combining the last two observations we see that the contact process is "self–dual":

(4) $P(\xi^A_t \cap B \neq \phi) = P(\xi^B_t \cap A \neq \phi).$

If we let ξ^1_t denote the contact process starting from $\xi^1_0 = \mathbb{Z}^d$, and set $A = \mathbb{Z}^d$ in (4), we get

(5) $$P(\xi_t^1 \cap B \neq \phi) = P(\xi_t^B \neq \phi).$$

Since ϕ is an absorbing set, the right—hand side decreases to a limit as $t \to \infty$. In Section 2 we observed that it is possible to write every probability of the form $P(\xi_t^1(x_1) = i_1,...,\xi_t^1(x_k) = i_k)$, where $x_1,...,x_k \in Z^d$ and $i_1,...,i_k \in \{0,1\}$ (and $\xi_t^1(x_i) = 1$ if $x_i \in \xi_t^1$), in terms of the $P(\xi_t^1 \cap B \neq \phi)$ with $B \subset \{x_1, ...,x_k\}$. It follows from the last two observations that the finite dimensional distributions of ξ_t^1 converge to those of a limit ξ_∞^1 which is a stationary distribution for the contact process.

The analysis above applies to all values of λ. If $P(\xi_t^0 \neq \phi \text{ for all } t) = 0$ then

$$P(0 \in \xi_\infty^1) = \lim_{t \to \infty} P(0 \in \xi_t^1) = \lim_{t \to \infty} P(\xi_t^0 \neq \phi) = 0$$

so $\xi_\infty^1 = \phi$ with probability 1. It is easy to see that in this case there are no stationary distributions other than the trivial one; δ_ϕ, the point mass on ϕ. (Proof: if ν is a stationary distribution, $\xi_t^\nu \subset \xi_t^1$ and the right—hand side $\to \phi$ as $t \to \infty$.) When $\lambda > \lambda_c$ the last argument shows $P(0 \in \xi_\infty^1) > 0$, so ξ_∞^1 is a nontrivial stationary distribution. The next result shows that all stationary distributions are a convex combination of δ_ϕ and ξ_∞^1. Although this result should be true in any d, we will prove it here for d = 1.

(6) Theorem. Let $\tau^A = \inf\{ t : \xi_t^A = \phi \}$. If $\lambda > \lambda_c$ then as $t \to \infty$

$$\xi_t^A \Rightarrow P(\tau^A < \infty)\delta_\phi + P(\tau^A = \infty)\xi_\infty^1.$$

Proof: On $\{\tau^A < \infty\}$, $\xi_t^A = \phi$ for all t sufficiently large, so what we need to show is that on $\{ \tau^A = \infty \}$, ξ_t^A looks like ξ_t^1. First consider the case $A = \{0\}$. From (5) and (6) in the last section, it follows that

$$\xi_t^0 = \xi_t^1 \cap [\ell_t^0, r_t^0]$$

where $\ell_t^0 = \inf \xi_t^0$ and $r_t^0 = \sup \xi_t^0$, and if $\ell_t = \inf \xi_t^{[0,\infty)}$, $r_t = \sup \xi_t^{(-\infty,0]}$ then ℓ_t^0
$= \ell_t$ and $r_t^0 = r_t$ on $\{\xi_t^0 \neq \phi\}$. As $t \to \infty$, $\ell_t/t \to -\alpha(\lambda) < 0$ and $r_t/t \to \alpha(\lambda) > 0$ so
on $\{\tau^0 = \infty\}$, ξ_t^0 agrees with ξ_t^Z on a linearly growing set.

From the last observation the desired result (for A = {0}) follows easily (see
Durrett (1980), p.902–903 for more details). To extend the result to a general A, we
use a "restart argument" or in less technical terms "if at first you don't succeed, try,
try again." Pick $y \in A$. (Say to minimize $|y-1/3|$.) If $\xi_t^y \neq \phi$ for all t, the last
argument applies. If $\tau^y = \inf\{ t : \xi_t^y = \phi \} < \infty$ and $\xi^A(\tau^y) = \phi$, then we are done
since the process has died out. If $\tau^y < \infty$, and $\xi^A(\tau^y) \neq \phi$, then we pick another
particle and try again. After at most a geometric number of trials, either $\xi_t^A = \phi$ or
we find a process that lives forever and agrees with ξ_t^1 on a linearly growing set. In
either case we have what we want, so the proof is complete. For more details see
Durrett (1980), p.903–904 this time.

The proof of (6) shows that on $\Omega_\infty = \{\xi_t^0 \neq \phi$ for all t$\}$ we have

$$\xi_t^0 = \xi_t^1 \cap [\ell_t, r_t] \approx \xi_\infty^1 \cap [-\alpha(\lambda)t, \alpha(\lambda)t].$$

If we revert to coordinate notation, setting $\xi_\infty^1(x) = 1$ if $x \in \xi_\infty^1$ and $= 0$ otherwise, then
$\xi_\infty^1(x)$, $x \in Z$ is a stationary sequence which is ergodic. (See Holley (1972), p.1967 and
Durrett (1980), p. 898, or better still, try to prove this yourself. It is not hard.) Based
on this, it should not be hard to believe, and is not difficult to prove:

(7) Theorem. Let $\lambda > \lambda_c$. If $\rho(\lambda) = P(0 \in \xi^1_\infty)$ then as $t \to \infty$

$$|\xi^0_t|/t \to 2\alpha(\lambda)\rho(\lambda)1_{\Omega_\infty} \quad \text{in } L^1.$$

To see why the limit is the right–hand side, observe that on Ω_∞, $[\ell^0_t, r^0_t] = [\ell_t, r_t]$ has width $\sim 2\alpha(\lambda)t$, and a fraction $\rho(\lambda)$ of the sites are in ξ^1_∞. (7) is proved in Durrett (1980), p.897–899. The reader probably expected to see almost sure convergence asserted in the conclusion. The stronger conclusion is true, but a proof had to wait until Durrett and Griffeath (1983), and the reader will have to wait until Section 10b. Since we will prove the stronger result later, we will not prove (7) now.

The last few results have considered what happens when $\lambda > \lambda_c$. The results are trivial when $\Omega_\infty = \{\xi^0_t \neq \phi \text{ for all t}\}$ has probability 0, so it is natural to ask if we have all the cases covered, or "Is $P(\Omega_\infty) = 0$ when $\lambda = \lambda_c$?" By analogy with what happens in a critical branching process and the biased voter, we expect the answer to be yes, but this has turned out to be a very difficult question. [If I were P. Erdös I would offer \$1000 for the solution of this problem. I'm not, and anyway, the reader would probably find it more profitable to take a part–time job at McDonald's!] All that we have been able to prove is that $\alpha(\lambda_c) = 0$ (see Section 10a), and that if $P(\Omega_\infty) > 0$ at λ_c, then on Ω_∞, r^0_t and ℓ^0_t return to 0 infinitely often (see Durrett (1984), Section 4), but $r^0_t - \ell^0_t \to \infty$ almost surely. The last state of affairs seems ridiculous, but as far as I can tell, does not contradict anything.

We close this section by giving a lower bound on the survival probability at λ_c which improves a result of Griffeath (1981).

(8) Theorem. If $\lambda = \lambda_c$ then as $t \to \infty$, $t^{1/2} P(\xi^0_t \neq \phi) \to \infty$.

Remark: For comparison purposes note that in a branching process the last probability $\sim C/t$ as $t \to \infty$. Combining numerical results from Brower, Furman, and Moshe (1978) with scaling relationships (12) and (18) from Grassberger and de la Torre (1979), gives

$$P(\ \xi_t^0 \neq \phi\ ) \approx t^{-.161} \text{ as } t \to \infty,$$

so we are far from the right result. However, (8) does show that the critical exponent δ defined by

$$\lim_{t \to \infty} \log P(\ \xi_t^0 \neq \phi\ )/ \log t = -\delta$$

does not take its mean field (i.e., branching process) value, $\delta = 1$.

Proof: Consider the contact process ξ_t^n starting from $[-n\epsilon, n\epsilon]$ occupied, let $r_t^n = \sup \xi_t^n$, and let $\ell_t^n = \inf \xi_t^n$. If we let $r_t = \sup \xi_t^{(-\infty,0]}$ and recall $\alpha(\lambda_c) = 0$ and $r_t/t \to 0$ a.s., an easy argument shows

$$\sup_{s \leq t} |r_s|/t \to 0 \text{ a.s.}$$

From this, it follows that the random processes $r^n(nt)/n$ and $\ell_t^n(nt)/n$ $0 \leq t \leq 1$ converge to functions that are constant at ϵ and $-\epsilon$. A coupling result, similar to (6) in the last section, shows

$$\tau^n \equiv \inf\{\ t : \xi_t^n = \phi\ \} = \inf\{\ t : \ell_t^n > r_t^n\ \},$$

so

$$P(\text{ there is a path from } [-n\epsilon,n\epsilon]\times\{0\} \text{ to } \mathbb{Z}\times\{n\} \text{ in } (-2n\epsilon,2n\epsilon)\times[0,n]) \to 1.$$

The last fact implies that, with a probability that approaches 1, there is a point m in $[-2n\epsilon,2n\epsilon]$ so that $m \in \xi^n_{n/2}$, and the contact process starting at m at time $n/2$ survives until time n. Using the duality equation (4), the probability of the last event for any fixed m is at most $P(\xi^0_{n/2} \neq \phi)^2$. So summing over m gives

$$\liminf_{n\to\infty} 4\epsilon n\, P(\xi^0_{n/2} \neq \phi)^2 \geq 1.$$

The last result gives (8) with ∞ replaced by $1/\sqrt{2\epsilon}$, but $\epsilon > 0$ is arbitrary, so the proof is complete.

5 One Dimensional Discrete Time Models

5a Oriented Percolation

In this and the next two sections, we will be concerned with discrete time models in which the state at time n is $\xi_n \subset \mathbb{Z}$. The simplest of these systems is oriented bond percolation, which is constructed as follows: if $(m,n) \in V \equiv \{ (m,n) \in \mathbb{Z}^2 : m + n$ is even$\}$, there is an oriented arc from (m,n) to $(m+1,n+1)$ and from (m,n) to $(m-1,n+1)$. Each arc, also called a bond, is independently designated as open with probability p and closed with probability $1-p$. We think of open bonds as air spaces which permit the passage of a fluid through the bond in the direction of the orientation. With this in mind, we make the following definitions:

$x \rightarrow y$ (y can be reached from x) if there is an open path from x to y; that is, there is a sequence $x_0 = x, \dots x_j = y$ of points in V such that for each $1 \le i \le j$ the bond from x_{i-1} to x_i is open

$C_{(0,0)}$ (the cluster containing $(0,0)$) is $\{ x : (0,0) \rightarrow x \}$

$\xi_n^0 = \{ m : (0,0) \rightarrow (m,n) \}$

$C_{(0,0)}$ is the set of sites that will be wet by a source of fluid at 0. ξ_n^0 is the collection of wet sites on level n.

When this process was introduced by Broadbent and Hammersley in (1957), they worked on $\mathbb{Z}^2$, and had bonds connecting z to $z + (1,0)$ and to $z + (0,1)$. To get a process that evolves in time, we have rotated the picture by 45° (and multiplied it by $\sqrt{2}$). The resulting process is a subset of the even integers at even times and a subset of the odd integers at odd times. The last feature is a little awkward, but it is simpler than other alternatives we know of (e.g., having bonds from (m,n) to (m,n+1) and from (m,n) to (m+1,n+1) — which is not spatially symmetric). In any case, the reader is stuck with this approach for the next five sections.

The first observation to be made about the model is that if p is too small then ξ_n^0 dies out, that is, $\xi_n^0 = \phi$ if n is large. To see this, observe that there are 2^n paths from (0,0) to "level n" = { (m,n) : m $\in$ Z }. Each path is open with probability p^n, so the expected number of open paths is $2^n p^n$. If $p < 1/2$ then

$$P(\ \xi_n^0 \neq \phi\ ) \leq 2^n p^n \to 0 \text{ as } n \to \infty,$$

so { $\xi_n^0 \neq \phi$ } $\downarrow \phi$ as n $\uparrow \infty$. If we let

$$p_c = \inf \{\ p : P(\ \xi_n^0 \neq \phi \text{ for all n }) > 0\ \},$$

then it follows from the last conclusion that $p_c \geq 1/2$.

With some work this lower bound can be improved considerably (see Durrett (1984a), Section 6), but that and many other properties of oriented percolation are explained in the paper cited, so we will content ourselves here to show $p_c < 1$. The method for proving this is called a "contour argument." Let

$$A = \{\ 0, -2, -4, \ldots -2N\ \}$$

$$C = \{\ z : \text{there is a } y \in A \text{ so that } (y,0) \to z\ \}$$

$$D = \{\ (a,b) \in \mathbb{R}^2 : |a| + |b| \leq 1\ \}$$

$$W = \cup_{z \in C} (z + D),$$

where $z + D = \{ z+w : w \in D \}$. D is for diamond. W is for wet region. In words, we have enlarged the cluster C to be a solid blob by replacing each point $z \in C$ by a diamond centered at that point. If $|C| < \infty$, let Γ be the boundary of the unbounded component of $(\mathbb{R} \times (-1,\infty)) - W$ and orient the boundary in such a way that the segment from $(0,-1)$ to $(1,0)$ (which is always present) is oriented in the direction indicated. Γ is called the contour associated with C. (See Figure 5.1.) In words, Γ is the "exterior boundary" of W.

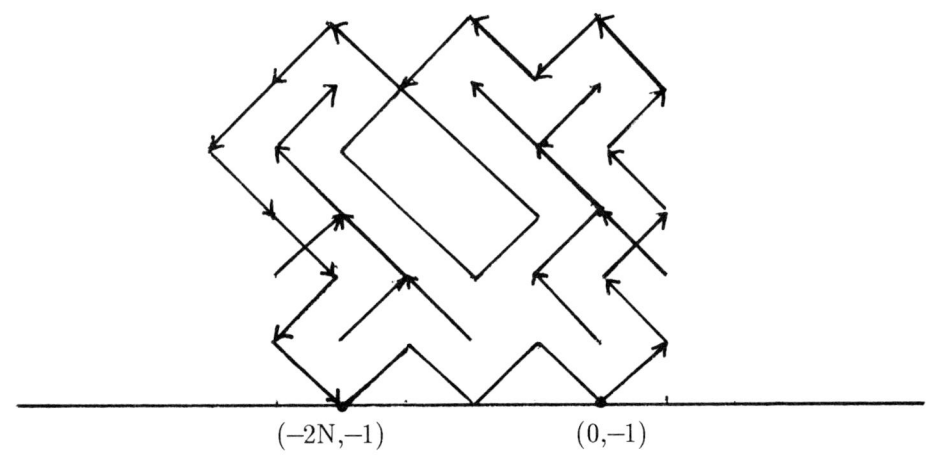

$(-2N,-1)$ $(0,-1)$

Figure 5.1

The key to the contour method is the observation

$$P(|C| < \infty) = P(\Gamma \text{ exists }) \leq E(\# \text{ of contours }).$$

To estimate the right–hand side we observe that:

(i) since the contour never passes through the same arc twice, there are at most 3^{n-1} contours of length n, and

(ii) a contour of length n must cut at least n/2 closed bonds.

To prove (i), we observe that the first segment is always $(0,-1) \to (1,0)$, and after that there are at most three choices at each stage, since we cannot use the bond we just traversed. To prove (ii), we look at Figure 5.1 (there the bonds drawn are open and all the others are closed), and note that the segments that go to the left (i.e., the x coordinate decreases in the direction of the orientation) must cut closed bonds, but those that go to the right need not. (Look at the two bonds that cross the contour. They are open but the fluid cannot go through them because of the orientation.) If we let n_ℓ and n_r be the number of segments of these two types then $n_\ell + n_r = n$, and since the contour starts at $(0,-1)$ and ends at $(-2N,-1)$, $n_\ell - n_r = 2N$. Adding the last two equations gives $2n_\ell = n + 2N$, proving (ii).

With (i) and (ii) established, the rest is easy. The shortest possible contour has length 2N+4, so

$$E(\ \# \text{ of contours }) \leq \sum_{n=2N+4}^{\infty} 3^{n-1}(1-p)^{n/2}$$

If $p > 8/9$, then the right–hand side $\to 0$ as $N \to \infty$, and it follows that for large N, $P(\ |C| = \infty\ ) > 0$. To get from this to $P(\ |C_{(0,0)}| = \infty) > 0$, observe that

$$P(\ |C| = \infty\ ) \leq (N+1)\ P(\ |C_{(0,0)}| = \infty).$$

Since if the event on the left happens, the cluster containing one of the points $(0,0)$, $(-2,0)$, ... $(-2N,0)$ must be infinite. Note that in the argument above, we could have taken $N = 0$; i.e., $A = \{0\}$, but this would have given us a worse bound on p_c and would not give us (2) below.

The bound provided by the last argument is very crude, but it is not far from the best known result; $p_c < .84$. (The improvement comes from getting a better bound on the number of contours. See Durrett (1984a) p.1028 for a proof.) It is an open problem to find a sequence of upper bounds that can be proved to converge to p_c. Even if a sequence of upper bounds were known, it seems likely that the sequence would converge slowly, like the lower bounds given in Section 6 of Durrett (1984a), and not provide a feasible way of computing p_c. This is a situation that will occur throughout the developments below. In most cases we do not have good bounds on the critical value, and we turn to the physics literature or resort to computer simulations to get an idea of what the critical value is.

The critical value of oriented percolation has been estimated by a number of nonrigorous methods: series expansion (Blease (1977), Essam and De'Bell (1981)), Monte Carlo (Kertesz and Vicsek (1980), Dhar and Barma (1981)), and finite size scaling (Kinzel and Yeomans (1981)). A consensus has developed that $p_c \approx .6446$, although these papers do not agree on the value of the last digit. For a survey of oriented percolation from the physicist's point of view, see Kinzel (1983). On the next three pages we show some simulations which are meant to suggest that $p_c \approx .64$. As in the simulations of the contact process given in Chapter 4, things have been arranged so that ξ_n is increasing in p. In the first picture, $p = .6$ and the process is dying out. The second has $p = .64 \approx p_c$. In the last, $p = .68$ and the right edge has positive drift. In the next section we will explain that, as in the contact process, "edge speeds characterize p_c."

P = .60

P = .64

P = .68

In Chapter 1, we introduced a discrete time growth model $\xi_n \subset \mathbb{Z}^2$ which evolves according to the following rules:

(i) if $x \in \xi_n$ then $x \in \xi_{n+1}$,

(ii) if $x \notin \xi_n$ then $P(\, x \notin \xi_n \mid \xi_n \,) = (1-p)^{\,|\xi_n \cap N(x)|}$,

where $N(x) = \{\, y : |y-x| = 1 \,\}$ is the set of neighbors of x. If $\xi_0 = \{0\}$ then ξ_n is contained in $\{\, x : |x| \le n \,\}$. If (x,y) has $x + y = n$ then we can have $(x,y) \in \xi_n$, only if $(x-1,y) \in \xi_{n-1}$ and $(x-1,y) \to (x,y)$ is open, OR $(x,y-1) \in \xi_{n-1}$ and $(x,y-1) \to (x,y)$ is open. From this, it follows that $C = \{\, (x,y) : x,y \ge 0 \ (x,y) \in \xi_{x+y} \,\}$ is a copy of oriented percolation, in the orientation considered by Broadbent and Hammersley. So if $p > p_c$ then the limiting shape discussed in Section 1 will intersect $\{\, x : x_1 + x_2 = 1 \,\}$ in a nonempty interval. The converse is also true, and is left as an exercise for the reader. (Read the next section, and then generalize results from Section 4b.)

Our next task is to show that for the contact process $\lambda_c < \infty$, tying up a loose end from Chapter 4. This and the percolation construction we will use in Section 10a, are related to oriented site percolation, so we will now describe that model. Site percolation takes place on the graph V defined above, but this time each point (m,n) in V, also called a site, is independently designated as open with probability p, and closed with probability $1-p$. We think of open sites as air spaces in a rock, and with this in mind, we make the following definitions:

$x \to y$ (y can be reached from x) if there is an open path from x to y; that is, there is a sequence $x_0 = x$, x_1, ... $x_j = y$ of points in V so that all the x_i are open and for $1 \le i \le j$ x_i is $x_{i-1} + (1,1)$ or $x_{i-1} + (-1,1)$.

$$C_{(0,0)} = \{ x : (0,0) \to x \}$$

$$\xi_n^0 = \{ m : (m,n) \in C_{(0,0)} \}$$

$$p_c = \inf\{ p : P(\xi_n^0 \neq \phi \text{ for all n }) > 0 \}$$

The argument given for the bond model tells us $p_c \geq 1/2$, so we turn our attention to the problem of showing $p_c < 1$. The method for doing this is again the "contour argument." We begin by defining A, C, D, W, and Γ exactly as before. Since the definition of Γ is unchanged, the number of contours is $\leq 3^{n-1}$. When we estimate the probability a contour exists, things change. This time, (ii) is replaced by "in order for a contour of length n to exist, there must be at least n/4 closed sites adjacent to Γ."

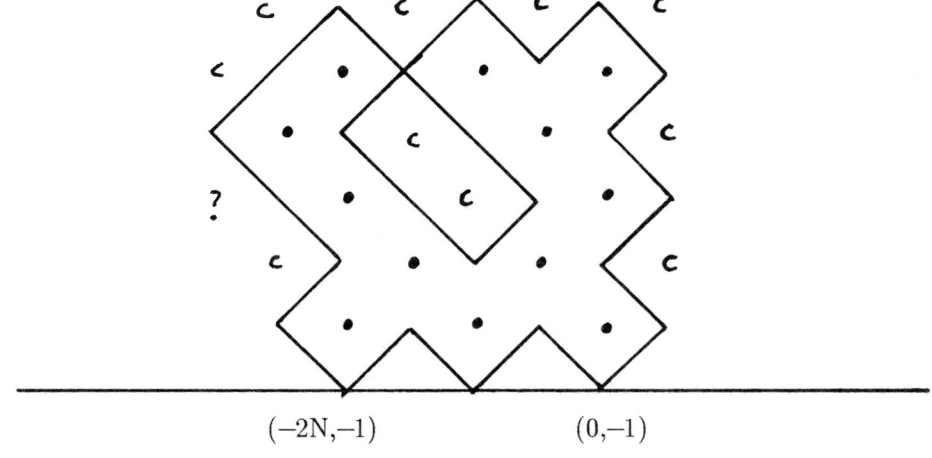

$$(-2N,-1) \qquad\qquad (0,-1)$$

Figure 5.2

To prove the claim in quotes, we need some definitions. If we stand at the midpoint of one of the segments that makes up Γ, and face in the direction of the orientation, then our left hand is in W and our right is in W^c. (See Figure 5.2.) The site closest to our right hand is called the site "associated" with the segment. Sites associated with segments that go to the left (i.e., the x coordinate decreases when we move in the direction of the orientation) must be closed, but those associated with segments that go to the right need not. If we let n_ℓ and n_r be the number of segments of these two types, then again $n_\ell + n_r = n$ and $n_\ell - n_r = 2N$, so $n_\ell = (n+2N)/2 \geq n/2$. However, a closed site may be associated with two boundary segments, so we only get that the number of closed sites $\geq n/4$.

The last conclusion is weaker than the one for bond percolation, but is still adequate for the contour argument. The shortest possible contour has length 2N+4 so

$$(1) \qquad\qquad \text{E(\# of contours)} \leq \sum_{n=2N+4}^{\infty} 3^{n-1}(1-p)^{n/4}.$$

If $p > 80/81$, then the last result implies

$$(2) \qquad\qquad \text{P(}|C| < \infty \text{)} \leq K_p \left(3(1-p)^{1/4} \right)^{2N}$$

where K_p is a constant that only depends on p. The last result shows $\text{P(}|C| < \infty \text{)} \to 0$ exponentially fast as $N \to \infty$, and we can conclude, as before, that $\text{P(}|C_{(0,0)}| = \infty \text{)} > 0$ and $p_c \leq 80/81$.

In Section 10a, we need to consider a site percolation process in which sites are independent only when $\|x-y\|_V > 1$, where $\|z\|_V = (|z_1| + |z_2|)/2$. (We divide by 2 so that the distance between adjacent points is 1.) If $x \in V$ there are eight $y \in V$ with $\|x-y\|_V = 1$, so if we use the contour argument above, we can conclude there are at least n/36 independent closed sites along the boundary of Γ. This gives us (1) with 4 replaced by 36, and we can conclude that if $p > 1 - 3^{-36}$ then

(3)
$$P(\ |C| < \infty\ ) \le K_p\ (\ 3(1-p)^{1/36})^{2N},$$

where K_p is a (new) constant which only depends on p. The bound that (3) gives on p_c is clearly ridiculous, but we will see in Chapter 10 that (3) is very useful. When combined with the "renormalized bond construction" in Section 10a, it allows us to conclude that the corresponding result holds for the contact process for all $\lambda > \lambda_c$:

(4)
$$P(\ \xi_t^{\{0,...N\}} = \phi \text{ for some } t\ ) \le C\ e^{-\gamma N}.$$

Recall $A = \{0, -2, ... -2N\ \}$, and $C = \{\ z : \text{there is a } y \in A \text{ so that } (y,0) \to z\ \}$.

Our last task in this section is to show that for the contact process $\lambda_c < \infty$, tying up a loose end from Chapter 4. We will do this by following Harris' original (1974) approach. We will show that if δ is small, and λ is large, then the contact process observed at times $0,\delta,2\delta,...$ dominates oriented site percolation with $p > 80/81 \ge p_c$. The proof is simple, but the bound on λ_c is ridiculous. Readers who have been through Harris' proof will notice that the proof below is simpler and improves his bound by a factor of 8 (or more).

Let $\delta > 0$ and call a site $(m,n) \in V$ open if (i) there is no death at m between time $(n-1)\delta$ and $(n+1)\delta$, and (ii) there is a birth from $m \to m+1$ and from $m \to m-1$ between time $n\delta$ and $(n+1)\delta$. It is easy to see that with these definitions the sites are independent, and if the oriented site model on V percolates, then the contact process lives forever. (See Figure 5.3. In the realization drawn there $(0,0)$, $(-1,1)$, and $(0,2)$ are open in the site percolation process, and there is a path in the graphical representation

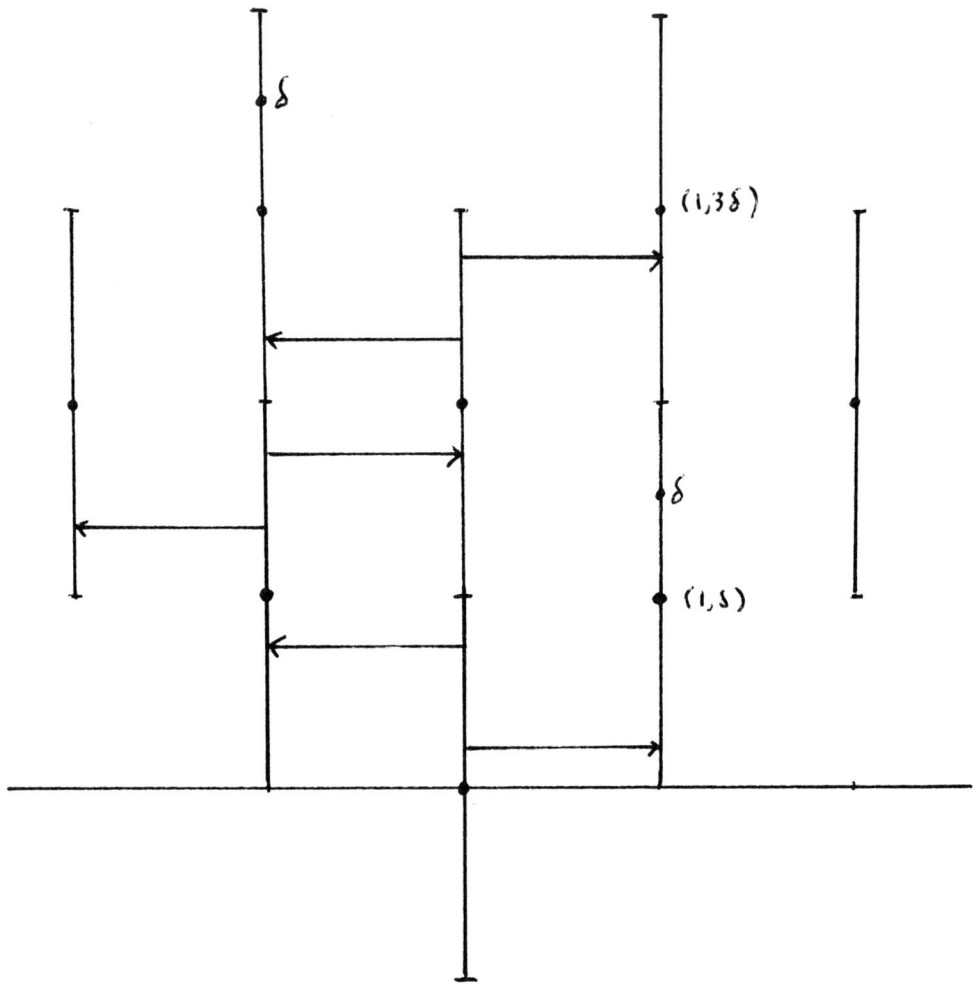

Figure 5.3

from $(0,0)$ to $(1,4\delta)$.) To estimate the probability a site is closed, we observe that

$$P(\text{ no death in } [0,2\delta]) = e^{-2\delta} \geq 1 - 2\delta,$$

and

$$P(\text{ birth from } m \to m+1 \text{ in } [0,\delta]) = 1 - e^{-\lambda\delta},$$

so

$$P(\text{ closed }) \leq 2\delta + 2e^{-\lambda\delta}.$$

We want the last quantity $< 1/81$. Picking $\delta = a/162$ forces us to take $2e^{-\lambda a/162} < (1-a)/81$ or $\lambda > (162/a)\ln(162/(1-a))$. When $a = 7/8$, this gives us $\lambda_c < 1328$. The last result is not very good when compared with Holley and Liggett's (1978) bound $\lambda_c \leq 2$, but the proof here is much simpler.

5b Phase Diagram for Two Neighbor Systems

In this section we will consider the set of discrete time processes ξ_n whose evolution satisfies: (i) $P(x \in \xi_{n+1} \mid \xi_n) = f(\mid \xi_n \cap \{x-1,x+1\} \mid)$, and (ii) given ξ_n the events $\{ x \in \xi_{n+1} \}$ are independent. Oriented bond and site percolation discussed in the last section are special cases. Here, as in the discussion of those models, we will take ξ_n to be a subset of the even integers for even n, and a subset of the odd integers for odd n. For otherwise, the model would split into two independent subsystems.

The inspiration for this section comes from an article of Kinzel (1985), or more precisely from Figure 2 on page 232 of that paper.

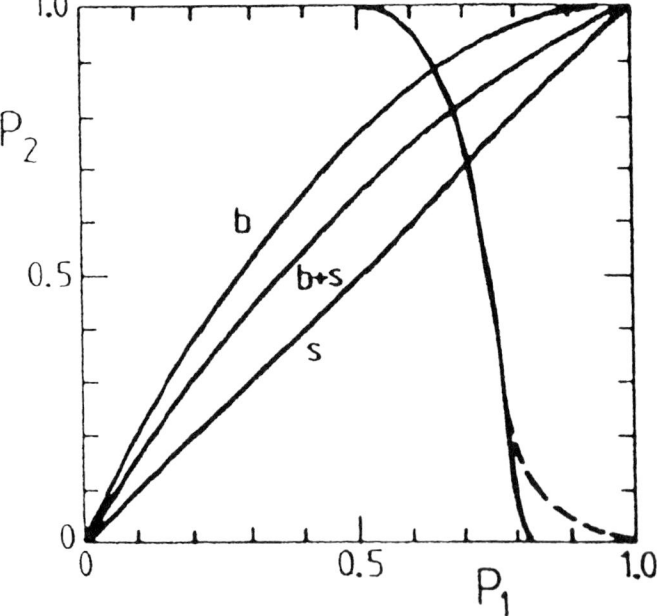

Figure 5.4

He considers the models generated by the rules above with $f(0) = 0$, $f(1) = p_1$, and $f(2)$ = p_2. The curve labelled b in that picture corresponds to bond percolation ($p_1 = p$, p_2 = $2p-p^2$). The line labelled s corresponds to site percolation ($p_1 = p_2 = p$). By considering a model in which sites are open with probability α, and bonds are open with probability β, one can construct rules with $p_1 = \alpha\beta$ and $p_2 = \alpha(2\beta-\beta^2)$. These rules lie between the curve marked b and the line marked s. If one takes $\alpha = \beta$ then the curve marked b+s results.

The rules described in the last paragraph are the only ones that can be achieved using a percolation structure with independent components. If we allow the state of $(m-1,n) \to (m,n+1)$ and $(m+1,n) \to (m,n+1)$ to be dependent (noting that this is consistent with (ii)), a slightly larger class of rules can be generated. Suppose we make

$(m-1,n) \to (m,n+1)$	$(m+1,n) \to (m,n+1)$	with prob.
open	open	a
closed	open	b
open	closed	b
closed	closed	$1-a-2b$

where bonds that END at different sites are independent. If we define a process by setting

$$\xi_n^x = \{\, y : (x,0) \to (y,n) \,\},$$

where "$(x,0) \to (y,n)$" is short for "there is an open path from $(x,0)$ to (y,n)" (the last phrase was explained in Section 5a), and

$$\xi_n^A = \bigcup_{x \in A} \xi_n^x,$$

then the resulting system has the form given above with $f(1) = a + b$, and $f(2) = a + 2b$. Taking into account the fact that we must have $a \geq 0$, $b \geq 0$, and $a + 2b \leq 1$, it follows that one of the processes we are considering can be represented in this way if and only if $f(1) \leq f(2) \leq 2f(1)$; or geometrically, if (p_1, p_2) lines in the triangle with vertices $(0,0)$, $(.5,1)$, and $(1,1)$.

The processes with $a = p$ and $b = 0$ have $f(1) = f(2) = p$, and correspond to site percolation. The models along the top edge of the square with $f(1) = p$ and $f(2) = 1$ are closely related to the biased voter model studied in Chapter 3. Drawing a picture:

$$
\begin{array}{llcccccccccc}
\text{time n} & \quad & 0 & 0 & 1 & 1 & 1 & 1 & 1 & 1 & 0 & 0 \\
\text{time n+1} & \quad & 0 & ? & 1 & 1 & 1 & 1 & 1 & ? & 0 &
\end{array}
$$

($? = 1$ with probability p and $? = 0$ with probability $1-p$), shows that if we let ξ_n^0 denote the system starting from $\xi_0^0 = \{\, 0\, \}$, then until $\tau = \inf \{\, n : \xi_n^0 = \phi\, \}$, ξ_n^0 consists of a solid block of sites $\{\ell_n^0, \ell_n^0 + 2, \ldots\ r_n^0 - 2, r_n^0\, \}$. A little more thought shows that if we let L_n and R_n be independent random walks with

$$
\begin{array}{ll}
P(L_{n+1} = L_n - 1) = p & P(\, L_{n=1} = L_n + 1) = 1 - p \\
P(R_{n+1} = R_n + 1) = p & P(\, R_{n+1} = R_n - 1) = 1 - p,
\end{array}
$$

then τ has the same distribution as $\sigma = \inf \{\, n : L_n > R_n\, \}$ and $\{\, (\ell_n^0, r_n^0) : n < \tau\, \}$ has the same distribution as $\{\, (L_n, R_n) : n < \sigma\, \}$. When $\ell_n^0 < r_n^0$, it is easy to see that the increments $\ell_{n+1}^0 - \ell_n^0$ and $r_{n+1}^0 - r_n^0$ are independent. To check this when $\ell_n^0 = r_n^0 = m$, we observe

$$
\{\, \ell_{n+1}^0 = m-1\, \} = \{\, (m,n) \to (m-1, n+1) \text{ is open}\, \},
$$

and

$$\{ \, r^0_{n+1} = m+1 \, \} = \{ \, (m,n) \to (m+1,n+1) \text{ is open } \}.$$

Since $E(R_1) = 2p-1$, and $E(L_1) = -(2p-1)$, it follows that in the class of models with $p_1 = p$ and $p_2 = 1$, $P(\, \xi^0_n \neq \phi \text{ for all } n \,) > 0$ if and only if $p > 1/2$. Having identified the critical value for the models along the top edge, it is natural to make this a starting point for investigating the rest of the square. Let

$$\rho(p_1,p_2) = P(\, \xi_n \neq \phi \text{ for all } n \,)$$

when the parameters take on the indicated values, and let

$$p_{1,c}(p_2) = \inf \{ \, p : \rho(p,p_2) > 0 \, \},$$

the critical value of p_1 for fixed p_2.

The arguments above show that $p_{1,c}(1) = 1/2$. The solid line in Figure 5.4 gives the values that "finite size scaling" predicts (see Section IV of Kinzel (1985) for details) for the curve $(p_{1,c}(p_2),p_2)$. We will ignore for the moment the dashed line in Figure 5.4 ending at $(1,0)$ and, in fact, for the rest of this section we will concentrate on what happens above the diagonal, that is, when $p_1 \leq p_2$. In this case, the system is called "attractive" and has the following very nice property: if $A \subset B$ then $\xi^A_n \subset \xi^B_n$ for all n. (Recall that the percolation construction defines the process simultaneously for all starting states.) To prove the last result, we observe that by induction it suffices to prove the result when $n=1$, and this is a trivial consequence of the definitions.

Extending the reasoning used in the last paragraph, it is easy to see that if (p_1,p_2) and $(\overline{p}_1,\overline{p}_2)$ are two points that have $p_1 \leq \overline{p}_1, p_2$ and $p_2 \leq \overline{p}_2$, and A and B are two sets with A ⊂ B, then copies ξ_n^A and $\bar{\xi}_n^B$ of these processes starting from $\xi_n^A = A$

and $\bar{\xi}_n^B = B$ can be constructed on the same space in such a way that $\xi_n^A \subset \bar{\xi}_n^B$ for all

n. Again, by induction, it is enough to verify the result when n = 1. To prove the

result in this case, let U_x be independent random variables that are uniform on (0,1),

and set

$$\xi_1^A(x) = 1 \text{ if } U_x < f(\, \xi_0^A(x{-}1) + \xi_0^A(x{+}1) \,)$$

$$\bar{\xi}_1^B(x) = 1 \text{ if } U_x < f(\, \bar{\xi}_0^B(x{-}1) + \bar{\xi}_0^B(x{+}1) \,).$$

Since $\xi_0^A(x{-}1) + \xi_0^A(x{+}1) \le \bar{\xi}_0^B(x{-}1) + \bar{\xi}_0^B(x{+}1)$, and we have assumed $p_1 \le \bar{p}_1, \bar{p}_2$

and $p_2 \le \bar{p}_2$, it follows that $\xi_1^A(x) \le \xi_1^B(x)$. (The inequality $p_1 \le \bar{p}_2$ is needed for the

case $\xi_0^A(x{-}1) + \xi_0^A(x{+}1) = 1$, $\bar{\xi}_0^B(x{-}1) + \bar{\xi}_0^B(x{+}1) = 2$.)

From results in the last paragraph, it follows that if $p_1 \le p_2$ and $\rho(p_1,p_2) > 0$

then $\rho(\bar{p}_1,\bar{p}_2) > 0$ whenever $\bar{p}_1 \ge p_1$ and $\bar{p}_2 \ge p_2$. (The last inequality implies $\bar{p}_2 \ge p_1$.)

Consequently p → $p_{1,c}(p)$ is decreasing while the curve $(p_{1,c}(p_2),p_2)$ lies above the

diagonal. We have underlined the last phrase to point out that while the curve drawn

in Figure 5.4 is monotone, the last result does not allow us to prove this in the region

$p_1 \ge p_2$, since we cannot compare, for example, the system with $p_1 = .85$, $p_2 = .4$ with

the one having $\bar{p}_1 = .9$, $\bar{p}_2 = .5$. All we can conclude about the curve below the

diagonal is that if p_s is the critical value of site percolation ($\approx .706$, see Kinzel and

Yeomans (1981)), then for $p \le p_s$ we have $p_{1,c}(p) \ge p_s$. The reader should also note

that while the first sentence in this paragraph implies that if $p_2 \ge p_s$ and $p > p_{1,c}(p_2)$

then $\rho(p,p_2) > 0$, we do not know if this is true when $p_2 < p_s$, that is, in this region we

do not know there is just one critical value.

We will discuss the systems that lie below the diagonal, (i.e., the case $p_1 \ge p_2$)

in the last two sections of this chapter. For the rest of this section, we concentrate on

what happens for attractive systems (i.e., the case $p_1 \le p_2$). Here, the situation is much like the contact process described in Chapter 4. If we let $r_n = \sup \xi_n^{(-\infty,0]}$, where $\xi_n^{(-\infty,0]}$ indicates the process starting from $\{0, -2, -4, \ldots\}$, then

(1)
$$\lim_{n \to \infty} r_n/n = \alpha(p_1, p_2),$$

(2)
$$p_{1,c}(p_2) = \inf \{\, p : \alpha(p,p_2) > 0 \,\} = \sup \{\, p : \alpha(p,p_2) < 0 \,\}.$$

The proofs of these results are almost the same as in Chapter 4, so they are omitted. See Durrett (1984a) for proofs for oriented percolation, and Durrett (1980) and Durrett and Griffeath (1983) for help in generalizing to the current situation.

To construct a dual process $\tilde{\xi}_n$, we use the same definitions as for ξ_n, but apply them to a graphical representation in which

$(m,n) \to (m+1,n+1)$	$(m,n) \to (m-1,n+1)$	with prob.
open	open	a
closed	open	b
open	closed	b
closed	closed	$1-a-2b$

and bonds that BEGIN at different sites are independent. If $a = p$ and $b = 0$ so that the original process is site percolation, then the dual is called a coalescing branching process. A particle at x splits into two with probability p and dies with probability $1-p$. The word "coalescing" refers to the fact that if two particles give birth onto the same site, only one particle results, or we can imagine that the two particles coaleasce into one.

It is easy to check that

(3) $P(\xi_n^A \cap B \neq \phi) = P(\tilde{\xi}_n^B \cap A \neq \phi).$

This duality equation implies, as in Chapter 4, that if we let ξ_n^1 denote the process starting from $\xi_0^1 = \mathbb{Z}$, then as $n \to \infty$, ξ_n^1 decreases to a limit ξ_∞^1 . The limit is again a stationary distribution for the process, but this time

(4) $P(0 \in \xi_\infty^1) = P(\tilde{\xi}_n^0 \neq \phi \text{ for all n }).$

(4) leaves open the possibility that site percolation has two critical values:

$$p_e = \inf \{ p : \xi_\infty^1 \neq \delta_\phi \}$$

$$p_f = \inf \{ p : P(\xi_n^0 \neq \phi \text{ for all n }) > 0 \}$$

where e is for equilibrium, and f is for survival starting from finite sets. All (2) tells us is that $p_e = \tilde{p}_f$ (where the $\sim$ refers to the dual process). To prove that $p_e = p_f = \tilde{p}_e = \tilde{p}_f$, we observe that if $r_n = \sup \xi_n^{(-\infty,0]}$ and $\ell_n = \inf \tilde{\xi}_n^{[0,\infty)}$ then

$$P(r_n \geq m) = P(\ell_n \leq -m);$$

since the left–hand side is the probability of a path from $(-\infty,0]\times\{0\}$ to $[m,\infty)\times\{n\}$, and the right is the probability of a dual path from $[m,\infty)\times\{0\}$ to $(-\infty,0]\times\{n\}$. From this, it follows that $\alpha(p_1,p_2) = \tilde{\alpha}(p_1,p_2)$, and (1) implies all the critical values are equal.

With the facts in the last paragraph in hand, it is easy to prove limit theorems like those in Chapter 4 for the limiting behavior of ξ_n^A and $|\xi_n^0|$. In short, our knowledge about the supercritical region $\{ (p_1,p_2) : p_{1,c}(p_2) < p_1 \leq p_2 \}$, and the behavior of these models is almost exactly the same as that for the contact process.

When it comes to the critical line, $\{ (p_1,p_2) : p_{1,c}(p_2) = p_1 \leq p_2 \}$, we do not know anything about what happens, except at the endpoint $p_1 = 1/2$, $p_2 = 1$, where we know $P(\xi_n^0 \neq \phi$ for all n $) = 0$. (Since L_n and R_n defined above are mean zero random walks.) We believe that all the processes on the critical line die out, and that with the exception of the endpoint of the critical line ($p_1 = 1/2$, $p_2 = 1$) where $\xi_\infty^1 = \mathbb{Z}$, for a trivial reason, that they all have $\xi_\infty^1 = \delta_\phi$.

Probing the last situation more deeply leads to an open problem that is one of the most intriguing in the subject. Recall we have defined $\rho(p_1,p_2) = P(\xi_n^0 \neq \phi$ for all n). If $p_2 = 1$ then, by observing that $R_n - L_n$ is a random walk on the even integers that takes steps of size -2, 0, and 2 with probabilities $(1-p)^2$, $2p(1-p)$, and p^2, we see

$$\rho(p,1) = P(R_n-L_n \geq 0 \text{ for all n }) = 1 - (1-p)^2/p^2,$$

so

$$\rho(1/2 + \epsilon,1) = \frac{2\epsilon}{(1/2 + \epsilon)^2} \sim 8\epsilon \quad \text{as } \epsilon \to 0.$$

From the last equation, it follows that if we define a critical exponent β (which depends on p_2) by

$$\rho(p,p_2) \approx (p - p_{1,c}(p_2))^\beta \text{ as } p \downarrow p_{1,c}(p_2),$$

where $f(p) \approx g(p)$ means $(\log f(p))/(\log g(p)) \to 1$, then $\beta(1) = 1$. Almost nothing is known rigorously about $\beta(p)$ for $p < 1$. (A result of J.T. and L. Chayes (1986) implies that for bond or site percolation $\beta \leq 1$, but we do not know how to show $\beta(p_2) \leq 1$.) A remarkable conjecture (imported from the physics literature where it is considered to be an established fact) is that $\beta(p)$ is constant for $p < 1$, and hence is always equal to the value for bond or site percolation. Numerically the last value is $\beta = .276 \pm .014$ (see Dhar and Barma (1981), p.L6) but this value must be taken with a grain of salt, since on the same page they say $\beta = .240 \pm .006$. The first value is consistent with the value

for the contact process that comes from the work of Brower, Furman, and Moshe
(1978), and Grassberger and de la Torre (1979) mentioned at the end of Section 4a.
The same idea of "universality" which predicts $\beta(p)$ is constant for $p < 1$, predicts that
the contact process has the same value of β.

At first glance, the conjecture that $\beta(p)$ is independent of $p < 1$ might seem to
contradict the fact that $\beta(1) = 1$, since it is known that $\rho(p_1,p_2)$ is continuous in the
supercritical region. (See Theorem 11 in Griffeath (1981). The proof given there for the
contact process generalizes easily to the current setting.) However it is NOT valid to
interchange limits to conclude that $\beta(p) \to \beta(1)$ as $p \to 1$, and as Michael Fisher
explained to me after a talk I gave in his seminar: "a crossover occurs." The last phrase
means that there is a function $\epsilon(p)$ which $\to 0$ as $p \to 1$, so that if $0 < p - p_{1,c}(p_2)$
$\ll \epsilon(p_2)$ ($\ll$ meaning much less than) then $\rho(p,p_2) \approx (p - p_{1,c}(p_2)^\beta$, while if the
difference is $\gg \epsilon(p_2)$ then $\rho(p,p_2) \approx \rho(p,1)$. To prove something like this is hopeless,
since we do not know how to show $\rho(p_1,p_2) = 0$ on the critical line. A first step in
understanding the situation would be to look at the behavior of $p_{1,c}(1-\epsilon)$ as $\epsilon \to 0$. It is
not hard to show $p_{1,c}(1-\epsilon) \geq (1+\epsilon)/2$ (by showing $\alpha(1/2 + \epsilon, 1-\epsilon) < 0$), but I think
that $p_{1,c}(1-\epsilon) - 1/2 \approx \epsilon^\theta$ where $\theta < 1$, and think that finding the right θ is an
interesting problem.

5c Back to Continuous Time

In the last section, we identified the discrete time models of the form

$$P(\ x \in \xi_{n+1} \mid \xi_n \ ) = f(\ |\xi_n \cap \{ \ x{-}1, x{+}1 \ \}| \ )$$

that could be constructed from a graphical representation. In this section, we will consider an analogous question in continuous time. We will suppose that the birth and death rates depend upon the number of occupied neighbors, that is, as $s \to 0$

(i) if $x \in \xi_t$, then $P(\ x \notin \xi_{t+s} \mid \xi_t \ ) \sim s \ \delta(\ |\xi_t \cap \{x{-}1,x{+}1\}| \ )$,

(ii) if $x \notin \xi_t$, then $P(\ x \in \xi_{t+s} \mid \xi_t \ ) \sim s \ \beta(\ |\xi_t \cap \{x{-}1,x{+}1\}| \ )$,

and we will see when the process can be constructed from a graphical representation.

We begin with a construction that produces the continuous time graphical representations as limits of ones with discrete time. Let $W_\epsilon = \{ \ (m,n\epsilon) : m,n \in \mathbb{Z} \ \}$, and consider W_ϵ as a graph with oriented bonds $(m{-}1,n\epsilon) \to (m,(n{+}1)\epsilon)$, $(m,n\epsilon) \to (m,(n{+}1)\epsilon)$, and $(m{+}1,n\epsilon) \to (m,(n{+}1)\epsilon)$. If we call the three types of bonds right, up, and left, then the graphical representation for Richardson's model can be obtained by letting the up bonds be open with probability 1, letting the left and right bonds be independently open with probability $\lambda\epsilon$, and letting $\epsilon \to 0$. To see this, note that the probability of no bond from m to m+1 before time t is

$$(1 - \lambda\epsilon)^{[t/\epsilon]+1} \ \to \ e^{-\lambda t} \quad \text{as } t \to \infty,$$

and with a little more work, one sees that the bonds from m to m+1 converge to a rate λ Poisson process. It is a little corny, but we would like to point out that in the limit, the bonds $(m,n\epsilon) \to (m,(n+1)\epsilon)$ become horizontal, so the random graph defined above converges to the graphical representation for Richardson's model.

Diverging from the order in which we introduced the processes, we consider the contact process next. In this case, we let the up bonds be open with probability $1-\epsilon$, the left and right bonds be open with probability $\lambda\epsilon$, and the state of the bonds be independent. The arguments for the Richardson's model model imply that the left and right bonds approach rate λ Poisson processes, as before, and the occurrence of vacant up bonds approaches a rate one Poisson process. In the limit, the lengths of the vacant up bonds approach 0, so to keep them from disappearing from the picture we mark their locations with δ's.

To get the graphical representation of the voter model as a limit, we need to use dependent bonds

left	up	right	probability
open	closed	closed	$\epsilon/2$
closed	open	closed	$1-\epsilon$
closed	closed	open	$\epsilon/2$

In the discrete time process, there is always exactly one open bond that ends at any point $(m,n\epsilon)$, so the state of m at time $n\epsilon$ is equal to that of a randomly chosen individual at time $(n-1)\epsilon$. That individual is m with probability $1-\epsilon$, and m+1 or m−1 with probability ϵ each. From this it should be clear that as $\epsilon \to 0$ the times at which the individual imitates a neighbor approaches a rate one Poisson process, and that when a neighbor is picked, m+1 and m−1 are chosen with equal probability.

As in the contact process, we mark closed up bonds by δ's. Notice that the second and third types of events produce $\longrightarrow\delta$'s and $\delta\!\!\leftarrow$'s in the limit, so what emerges as the limit of our random graphs is exactly the graphical representation for the voter model. The limiting procedure used to construct the voter model should make it clear that in the graphical representation when a $\delta\!\!\leftarrow$ happens, the δ occurs "just before" the arrow. The last observation should make the rules we used to define paths in Section 2a and Chapter 3 seem more natural.

The limiting procedure also leads easily to a description of all the possible graphical representations in continuous time. To do this, we observe that if we only allow dependence between bonds that end at the same point in W_ϵ, there are eight possibilities:

	left	up	right	probability
$\longrightarrow$	closed	open	closed	$1-c\epsilon$
	open	open	closed	$b(1)\epsilon$
$\longleftarrow$	closed	open	open	$b(1)\epsilon$
$\longrightarrow\!\!\leftarrow$	open	open	open	$b(2)\epsilon$
δ	closed	closed	closed	$d(0)\epsilon$
$\longrightarrow\delta$	open	closed	closed	$d(1)\epsilon$
$\delta\!\!\leftarrow$	closed	closed	open	$d(1)\epsilon$
$\longrightarrow\delta\!\!\leftarrow$	open	closed	open	$d(2)\epsilon$

where $c = 2b(1) + b(2) + d(0) + 2d(1) + d(2)$. In the table above, we have drawn the gadget in the graphical representation that corresponds to the limit of these bonds. Only the fourth and eighth lines are new. In the fourth, there will be a birth at m if either m−1 or m+1 is occupied, and in the eighth, there will be a death at m only if both m+1 and m−1 are vacant.

A little arithmetic shows that the birth and death rates given above are

$$\begin{aligned}
\beta(0) &= 0 \\
\beta(1) &= b(1)+b(2) \\
\beta(2) &= 2b(1)+b(2)
\end{aligned} \qquad\qquad
\begin{aligned}
\delta(0) &= d(0)+ 2d(1)+ d(2) \\
\delta(1) &= d(0)+ d(1) \\
\delta(2) &= d(0).
\end{aligned}$$

Since all the b's and d's must be nonnegative, there can be a graphical representation if and only if

$$b(1) = \beta(2) - \beta(1) \geq 0$$
$$b(2) = 2\beta(1) - \beta(2) \geq 0$$

$$d(0) = \delta(2) \geq 0$$
$$d(1) = \delta(1) - \delta(2) \geq 0$$
$$d(2) = \delta(0) - 2\delta(1) + \delta(2) \geq 0.$$

The second set of inequalities say that δ is decreasing and concave, while the first are usually written as $\beta(1) \leq \beta(2) \leq 2\beta(1)$.

If the death rates are constant, and we let $\beta(1) = \lambda$, then the last inequality says $\lambda \leq \beta(2) \leq 2\lambda$. When $\beta(2) = 2\lambda$, we have the "basic" contact process. At the other extreme, we have the "threshold" contact process, which has birth rate λ if at least one of the neighbors is occupied. We invite the reader to construct the dual for this model (the threshold contact process is the continuous time analogue of site percolation), and consider how its critical value is related to the basic contact process (the continuous time analogue of bond percolation). Answers can be found in Durrett and Griffeath (1983). We will have more to say about site versus bond percolation in Section 6b.

The inequalities derived in the last paragraph show that in continuous time, the assumption that the process comes from a graphical representation is a nontrivial restriction, but one that covers a number of interesting cases. As in discrete time, the theory for the processes considered above closely parallels that for the contact process, so we will not go into details.

Gray (1986) considered what happens when the inequalities above are weakened to $0 \leq \beta(1) \leq \beta(2)$, $\delta(2) \geq \delta(1) \geq \delta(0)$, the so-called attractive processes. These processes have the property that "more is better": if $A \subset B$ then processes ξ_t^A and ξ_t^B with $\xi_0^A = A$ and $\xi_0^B = B$ can be constructed on the same space, in such a way that $\xi_t^A \subset$

ξ_t^B for all t. The last property (take B = Z and A = ξ_s^Z) implies that ξ_t^Z decreases to a limit in the sense that P($\xi_t^Z \cap B \neq \phi$) $\downarrow$ a limit for all B, and it follows as in Sections 2a and 4a that the limit ξ_∞^Z is a stationary distribution.

For these systems, the arguments in Section 4a can be repeated to show

(1) the right edge r_t = sup $\xi_t^{(-\infty,0]}$ has an asymptotic speed; that is, $r_t/t \to \alpha(\lambda)$ a.s.,

and the coupling result

(2) if $0 \in A$ then $\xi_t^0 = [\ell_t^0, r_t^0] \cap \xi_t^Z$ on { $\xi_t^0 \neq \phi$ },

where ℓ_t^0 = inf ξ_t^0 and r_t^0 = sup ξ_t^0. With this in hand, to prove the "complete convergence theorem"

(3) $$\xi_t^A \Rightarrow \delta_\phi P(\tau^A < \infty) + \xi_\infty^Z P(\tau^A = \infty)$$

the only missing detail is to prove λ_c = inf{ $\lambda : \alpha(\lambda) > 0$ }. Looking back at Section 4a, one sees that to do this it is sufficient to prove that { $\lambda : \alpha(\lambda) = 0$ } is not an interval. In doing this in Section 4a, we used (7) which used the graphical representation in a crucial way.

The fact that some work is needed to fill this LITTLE gap can be established by noting that all of the other ingredients were available in Durrett (1980), but the solution had to wait until Gray's (1986) paper. Gray's proof is based on a graphical representation for attractive processes. One has for each site, a Poisson process with rate equal to the maximum birth or death rate, and a sequence of i.i.d. uniform random variables. We leave it to the reader to figure out what to do with these ingredients. The clever part of the proof is an inequality which says that if we stand at the right edge $r_t = \sup \xi_t^{(-\infty,0]}$, then the process to the left of us looks thicker than the process in equilibrium. To steal a line from Chung (1974) (see p.341), "the reader is urged to ponder over the intuitive meaning of the last result and judge for himself whether it is obvious or incredible." Remember all the sites to the right of the rightmost occupied site are vacant.

Gray's theorems do not require that the birth and death rates are of the form $f(|\xi \cap \{x-1, x+1\}|)$, but it is crucial that they just depend on the state of the nearest neighbors $x-1$ and $x+1$ (and the state of x itself). When the birth and death rates depend on $x-L$, ... $x+R$, where L or R > 1, edge speeds exist for attractive interactions, but the coupling result (2) breaks down and the game changes considerably. Durrett and Schonmann (1987) have shown, using the methods of Chapter 10, that for discrete time systems in $d = 1$ constructed from graphical representations, "edge speeds characterize the critical value," and all the theorems mentioned in the last section hold. We will return to the last topic in Section 10d. For a variety of technical reasons, Durrett and Schonmann are not able to prove their results in continuous time, and no one has a clue how to approach the problem for a general attractive process.

If proving things for a general attractive process is too frightening, the following example might be a good place to start. Suppose particles die at rate one and are born at rate λ if (x+1 and x+2) or (x-1 and x-2) are occupied (and at rate 2λ if all four points x+2,x+1,x-1, and x-2 are occupied). This process is a version of the contact process with "sexual reproduction." It is easy to show $\lambda_c < \infty$ (compare with the contact process on $2\mathbb{Z}$). Your mission is to prove that "edge speeds characterize the critical value" and/or that the complete convergence theorem holds for all $\lambda > \lambda_c$. As you can probably guess, we have no idea what happens in this model when $\lambda = \lambda_c$.

Some authors conjecture that the process survives at the critical value (see Durrett and Gray (1986)), but others say that this is only true in dimensions d $\geq$ 4 (see Grassberger (1982)). An open problem of some significance is to decide if the critical values defined by (a) survival of the process starting from a finite set, and (b) existence of a nontrivial stationary distribution, are the same.

5d Pascal's Triangle Mod 2

In this section, we return to discrete time, and continue to investigate the class of models introduced in Section 5b (using the notation introduced there), but this time we concentrate on what happens below the diagonal. We begin with the deterministic rule in the lower right corner $-$ $f(1) = 1$, $f(2) = 0$. If we let $\xi_n(x) = 1$ if $x \in \xi_n$ and $= 0$ otherwise, then the evolution can be written as:

$$\xi_{n+1}(x) = \xi_n(x+1) + \xi_n(x-1) \bmod 2.$$

To explain the name given in the title of this section, we observe that (i) if we let $N_0(0) = 1$, $N_0(x) = 0$ otherwise, and let

$$N_{n+1}(x) = N_n(x+1) + N_n(x-1),$$

then (if we ignore the 0's) the result is Pascal's triangle; and (ii) if we let $\xi_n^{\{0\}}(0) = 1$, and $\xi_n^{\{0\}}(x) = 0$ otherwise, then

(1)
$$\xi_n^{\{0\}}(x) = N_n(x) \bmod 2.$$

The best way to get a feeling for how $\xi_n^{\{0\}}$ behaves is to compute the behavior for a few steps. This is done in Figure 5.5. We have omitted the values of $\xi_n^{\{0\}}(x)$ when $n+x$ is odd or $|x| > n$, since these values are always 0. Periods are used to represent the other 0's to make the pattern more apparent. At times $2^n - 1$ with $n \geq 1$, $\xi_n^{\{0\}}$ is an interval of 2^n one's. Then at the next time, it consists of two one's separated by a

TIME:

```
 0                        1
 1                       1 1
 2                      1 . 1
 3                     1 1 1 1
 4                    1 . . . 1
 5                   1 1 . . 1 1
 6                  1 . 1 . 1 . 1
 7                 1 1 1 1 1 1 1 1
 8                1 . . . . . . . 1
 9               1 1 . . . . . . 1 1
10              1 . 1 . . . . . 1 . 1
11             1 1 1 1 . . . . 1 1 1 1
12            1 . . . 1 . . . 1 . . . 1
13           1 1 . . 1 1 . . 1 1 . . 1 1
14          1 . 1 . 1 . 1 . 1 . 1 . 1 . 1
15         1 1 1 1 1 1 1 1 1 1 1 1 1 1 1 1
16        1 . . . . . . . . . . . . . . . 1
```

Figure 5.5

distance of 2^n, and these one's start two copies of the original process that come together to make another interval of one's at time $2^{n+1} - 1$, and so on.

This "cellular automaton" has been studied in a number of papers. The pattern above is a "self-similar structure with fractal dimension $\log_2(3) \approx 1.5850$." The last phrase is a fancy way of saying that if you run the process for time 2t then you have 3 copies of the process run for time t (as long as $t = 2^n$). To see why this means the dimension is $\log_2(3)$, observe that if we double the side length of a cube in d dimensions then we have 2^d copies of the original and $\log_2(2^d) = d$.

Another phrase that is used in connection with this system is that "it is chaotic." (See Wolfram (1983), this is his rule 90.) This statement is much harder to explain but much more fun to unravel. Let $\nu_\theta =$ product measure with density θ, that is, the events { $\xi_0(x) = 1$ } are independent and have probability θ.

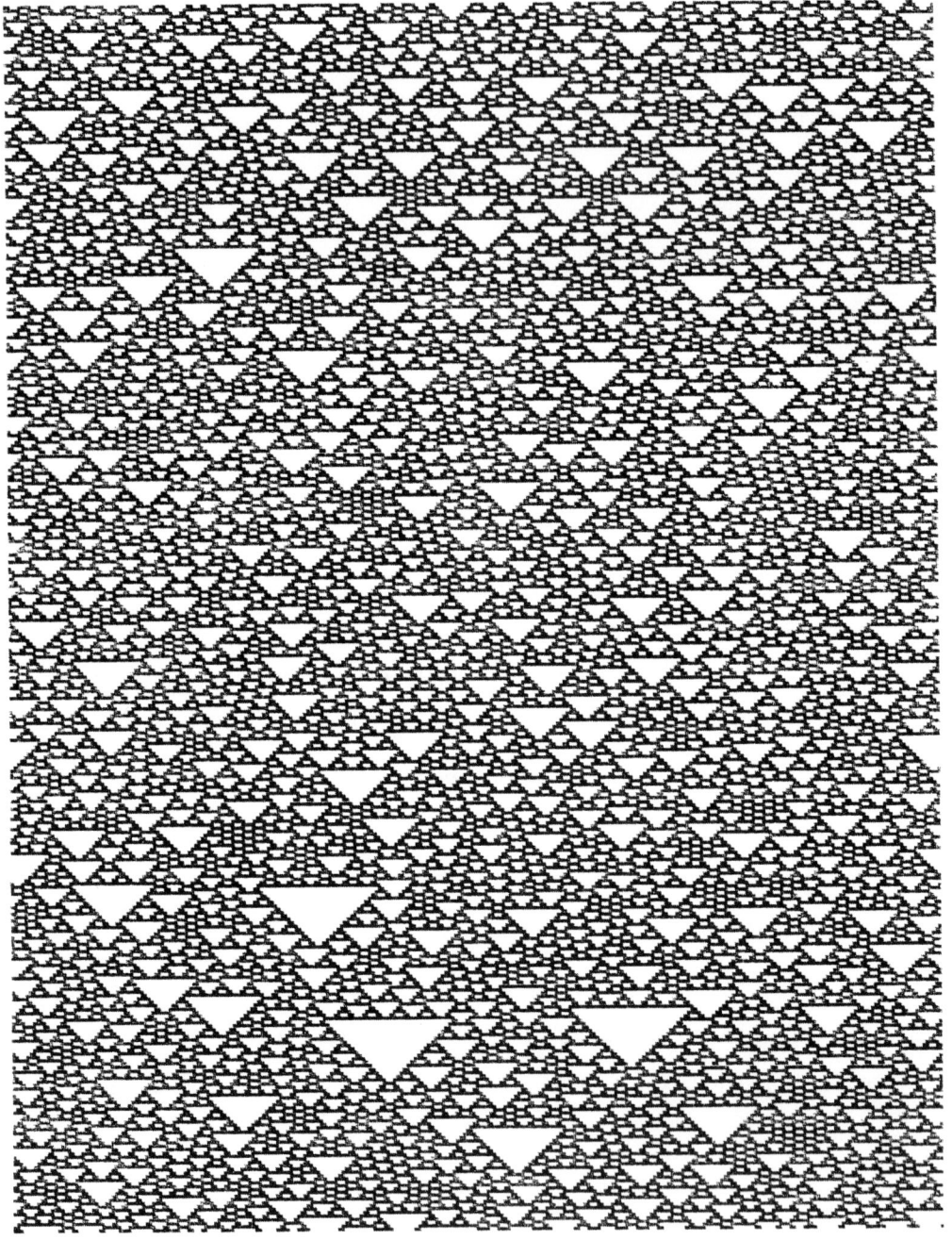

Figure 5.6

(2) $\nu_{1/2}$ is a stationary distribution for the process.

Proof: First, observe that if the initial configuration has distribution $\nu_{1/2}$ then

$$P(\ \xi_1(x) = 1\ ) = P(\ \xi_0(x+1) + \xi_0(x-1) \text{ is odd }) = 1/2.$$

To see that the sites in ξ_1 are independent, observe that conditioning on the values of $\xi_0(y)$ $y \leq x - 1$ shows that $\xi_1(x)$ is independent of $\xi_1(y)$ $y < x$.

Figure 5.6 shows the behavior of the process starting from $\nu_{1/2}$. Having seen that product measure with density $1/2$ is a stationary distribution, it is natural to ask what happens if we start with ν_θ, $\theta \neq 1/2$. In two cases the answer is trivial. If $\theta = 0$, that is, $\xi_0(x) \equiv 0$, then $\xi_n(x) \equiv 0$ for all n (so μ_0 is a trivial stationary distribution). If $\theta = 1$, that is, $\xi_0(x) \equiv 1$ then $\xi_1(x) \equiv 0$ and the system stays $\equiv 0$ for all $n \geq 1$. The next result describes the behavior in the interesting cases $0 < \theta < 1$.

(3) Theorem. Let $\tau^n \nu_\theta$ denote the distribution at time n when we start from ν_θ.
(i) If $\theta \notin \{0, 1/2, 1\}$, then $\tau^n \nu_\theta$ does not converge.
(ii) If $0 < \theta < 1$, then as $N \to \infty$

$$\frac{1}{N} \sum_{n=1}^{N} \tau^n \nu_\theta \Rightarrow \nu_{1/2},$$

where $\Rightarrow$ denotes convergence in distribution. (In this setting, convergence of finite dimensional distributions.)

Proof: This result was first proved by Miyamoto (1979), and later independently by Lind (1984). Since we found the second paper first, we will follow it. Being an ergodic theorist, Lind used Fourier analysis to prove the result. Experts in particle systems will recognize his approach as being that of Holley and Stroock (1979). In keeping with the "graphical representation" approach we have taken in Chapters 1–4, we will prove the result using Griffeath's (1979) idea of cancellative duality.

To formulate the duality, we go back to looking at ξ_n as a set valued process ($\xi_n = \{ x : \xi_n(x) = 1 \}$), and let ξ_n^A be the state at time n when $\xi_0^A = A$. The duality equation is

$$(4) \qquad\qquad P(\ |\xi_n^A \cap B| \text{ is odd }) = P(\ |\xi_n^B \cap A| \text{ is odd })$$

(as long as at least one of the sets A and B is finite). To prove this let $N_n(x)$ be Pascal's triangle (defined at the beginning of this section), and observe that if we draw an oriented graph with edges from (x,n) to (x+1,n+1), and from (x,n) to (x–1,n+1), then $N_n(x)$ is the number of paths from (0,0) to (x,n). This can be easily checked by induction, since the number of paths and $N_n(x)$ both satisfy: $N_{n+1}(x) = N_n(x–1) + N_n(x+1)$.

Generalizing $N_n(x)$ by setting $N_n^x(y) = N_n(y–x)$, and

$$N_n^A(y) = \sum_{x \in A} N_n^x(y),$$

we see that $N_n^A(y)$ is the number of paths from A x {0} up to (y,n). Reintroducing "coordinate notation." $\xi_n^A(y) = 1$ if $y \in A$, = 0 otherwise, we get a relationship that generalizes (1):

$$(5) \qquad\qquad \xi_n^A(y) = N_n^A(y) \text{ mod } 2.$$

The proof of (5) is trivial ($N_{n+1}^A(y) = N_n^A(y-1) + N_n^A(y+1)$ and "mod 2" commutes with addition), but it gives us exactly what we need to prove (4):

$$1_{(\, \xi_n^A \cap B \text{ is odd})} = (\, \sum_{x \in B} \xi_n^A(x) \,) \bmod 2 = (\, \sum_{x \in B} N_n^A(x) \,) \bmod 2$$

$$= 1_{(\, \# \text{ of paths from A x } \{0\} \text{ to B x } \{n\} \text{ is odd})}$$

$$= 1_{(\, \# \text{ of paths from B x } \{0\} \text{ to A x } \{n\} \text{ is odd})}$$

$$= 1_{(\, \xi_n^B \cap A \text{ is odd})}$$

(The restriction that at least one of the sets A and B is finite is needed to make the sums above finite.)

With (4) established, the proof of (3) is easy. Let ξ_n^θ denote the process starting from product measure with density θ. Using the duality equation (4) with $A = \xi_0^\theta$ and $B = \{0\}$, gives

$$(6) \qquad\qquad P\,(\, 0 \in \xi_n^\theta \,) = P\,(\, |\xi_n^{\{0\}} \cap \xi_0^\theta| \text{ is odd }).$$

If $|\xi_n^{\{0\}}| = k$, then the right–hand side is the probability $X_1 + X_2 \cdots + X_k$ is odd, where the X_i are independent with $P(X_i = 1) = \theta$ and $P(X_i = 0) = 1 - \theta$. If we let p_k be this probability, then it is clear that $p_1 = \theta$, and for $k \geq 1$

$$p_{k+1} = p_k(1-\theta) + (1-p_k)\theta.$$

Subtracting 1/2 from each side of the last equation gives

$$p_{k+1} - 1/2 = (p_k - 1/2)(1-\theta) + (1-p_k - 1/2)\theta = (p_k - 1/2)(1 - 2\theta).$$

So if $0 < \theta < 1$, $p_k \to 1/2$ exponentially fast as $n \to \infty$.

To prove (i) now, we let $n = 2^m$ in (6), and observe that $|\xi_n^{\{0\}}| = 2$, so

$$P(\, 0 \in \xi_n^{\theta}\,) = p_2 = 2\theta(1-\theta).$$

But at time $n-1$, $|\xi_n^{\{0\}}| = 2^m$, so

$$P(\, 0 \in \xi_n^{\theta}\,) = p(2^m) \approx 1/2$$

for $0 < \theta < 1$, and the sequence cannot converge unless $\theta = 1/2$ (or 0 or 1). To prove (ii) we begin by observing that $|\xi_n^{\{0\}}|$ is given by

1,

2,

2, 4,

2, 4, 4, 8,

2, 4, 4, 8, 4, 8, 8, 16,

2, 4, 4, 8, 4, 8, 8, 16, 4, 8, 8, 16, 8, 16, 16, 32, ...

where the repeating pattern is that for $k \geq 2$ the kth row (which gives the $|\xi_n^{\{0\}}|$ for $2^{k-1} \leq n < 2^k$) is 2 times the union of the first $k-1$ rows. More simply, if $k \geq 3$, the kth row is the $(k-1)$th row followed by 2 times that row. From the last observation, it follows immediately that as $N \to \infty$

(7) $$\frac{1}{N} |\{\, n \leq N : |\xi_n^{\{0\}}| \leq M \,\}| \to 0$$

for all $M < \infty$, and applying (6), that

$$\frac{1}{N} \sum_{n=1}^{N} P(\, 0 \in \xi_n^\theta \,) \to 1/2$$

for all $0 < \theta < 1$. To prove (ii), we have to show first that (7) holds when $\{0\}$ is replaced by any finite set B ($\neq \phi$), and second that if

(8)
$$\frac{1}{N} \sum_{n=1}^{N} P(\, |\xi_n^\theta \cap B| \text{ is odd }) \to 1/2$$

holds for any finite set B ($\neq \phi$) then (ii) holds. Both of these details are left to the reader. The second can be easily checked by hand — the probabilities in (8) determine the finite dimensional distributions. (A more sophisticated approach is to observe $\{0,1\}^{\mathbb{Z}}$ is a group under addition mod 2, and we have just checked that the Fourier transforms converge to that of $\nu_{1/2}$.) The first detail is much more difficult but the details are unpleasant, so we will take Lind's (1984) approach and say, with tongue in cheek, that this "can be easily deduced from properties of binomial coefficients modulo 2."

The last incantation completes the proof of (3). Having seen that $\nu_{1/2}$ and ν_0 are stationary distributions, it is natural to ask if there are any others. It is easy to see that the answer is "no" in the class of measures that are translation invariant and mixing. (Use the duality equation with $n = 2^m$ to conclude that μ must be a product measure.) Can the reader take this further, and show that the only translation invariant stationary distributions are of the form $\theta\, \delta_\phi + (1-\theta)\, \nu_{1/2}$? Of course, it would be even nicer to prove this without assuming translation invariance.

5e Cancellative systems

In this section, we consider a class of models that generalize the one considered in Section 5d. They are constructed from "graphical representations," but then we count the number of paths modulo 2. First, as in Section 5b, we construct a random graph in which

$(m-1,n) \to (m,n+1)$	$(m+1,n) \to (m,n+1)$	with prob
open	open	a
closed	open	b
open	closed	b
closed	closed	$1-a-2b$

and bonds that END at different sites are independent. Then, as in Section 5d, we let

$$N_n^x(y) = \text{the number of paths from } (x,0) \text{ to } (y,n)$$

$$N_n^A(y) = \sum_{x \in A} N_n^x(y)$$

$$\eta_n^A(y) = N_n^A(y) \bmod 2$$

A little arithmetic shows that the process just constructed has, in the notation of Section 5b, $f(1) = a + b$ and $f(2) = 2b$. In order for a, b, $1-a-2b \geq 0$, we must have $2f(1) - f(2) \geq 0$ and $f(1) + f(2)/2 \leq 1$. Geometrically, (p_1,p_2) must lie in the triangle with vertices $(0,0)$, $(.5,1)$, and $(1,0)$. This triangle sounds like the one in Section 5b, but there the last point is $(1,1)$ instead.

Since there is considerable overlap with the old examples, we will concentrate here on the models with $a = p$ and $b = 0$, which lie along the bottom of the square, outside the reach of the previous construction. A little thought shows that in these systems:

$$\eta^A_{n+1}(x) = (\ \eta^A_n(x+1) + \eta^A_n(x-1)\ ) \bmod 2 \qquad\qquad \text{with prob p}$$
$$= 0 \qquad\qquad \text{with prob 1-p}.$$

This is the drunk version of addition mod 2. With probability p, she writes down the right answer, and with probability 1−p, she forgets and writes down a 0. These systems are related to the dashed line in Figure 5.5. As Kinzel (1985) says in his paper: "The full line is the result of scaling systems with 10, 9, and 8 sites." "Presumably the ordered state of Fig. 3 (our Figure 5.5) is not dense enough to support a transition away from the deterministic limit, and the phase boundary behaves as shown by the dashed line in Fig. 2" (our Figure 5.4).

Kinzel's conclusion is based on the fact that

$$\frac{1}{N}\ \sum_{n=1}^{N}\ |\ \eta^{\{0\}}_n\ |\ \to 0 \text{ when } p = 1,$$

and the idea that "things will get worse when p < 1." The conclusion in quotation marks is wrong. The simulation in Figure 5.7 shows that when p = .98, the pretty pattern for p = 1 is destroyed. The mass executions that used to occur at times 2^m do not take place, and the system is thicker than when p = 1.

While the computer sometimes lies, this time it is telling the truth. Recently, Maury Bramson has figured out how to prove that

$$p_c = \inf\{\ p : P(\ \eta^{\{0\}}_n \neq \phi \text{ for all n }) > 0\ \} < 1.$$

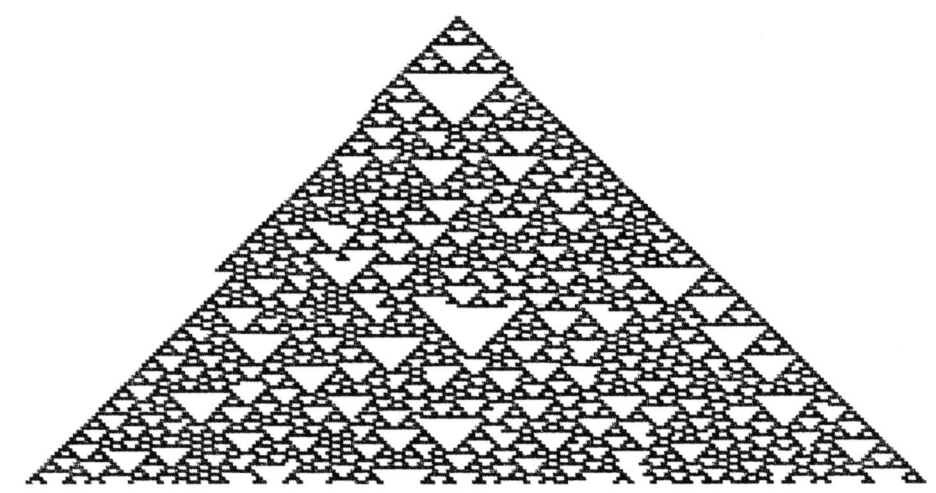

Figure 5.7

With the existence of the "phase transition" established, our next step is to investigate properties of the model using "cancellative duality." Consider the graphical representation that is obtained from the one above by reversing the direction of time. That is:

$(m,n) \rightarrow (m+1,n+1)$	$(m,n) \rightarrow (m-1,n+1)$	with prob
open	open	a
closed	open	b
open	closed	b
closed	closed	1−a−2b

and bonds that BEGIN at different sites are independent. If we let

$$\tilde{N}_n^x(y) = \text{the number of paths from } (x,0) \text{ to } (y,n)$$

$$\tilde{N}_n^A(y) = \sum_{x \in A} \tilde{N}_n^x(y)$$

$$\tilde{\eta}_n^A(y) = \tilde{N}_n^A(y) \bmod 2$$

then we get a process that is dual to η_n, in the sense that if we consider η_n and $\tilde{\eta}_n$ as set valued processes then

(1) $$P(\ |\eta_n^A \cap B| \text{ is odd }) = P(\ |A \cap \tilde{\eta}_n^B| \text{ is odd }).$$

The proof of the last equation is the same as that of (4) in the last section and is omitted.

When a = p and b = 0, the dual process $\tilde{\eta}_n$ is an annihilating branching process. A particle at x at time n branches into two particles put at x+1 and x−1 at time n+1 with probability p, and dies with no children with probability 1−p. The annihilation comes from the fact that if two particles give birth onto the same site, their two offspring annihilate each other and an empty site results.

When a = 0 and b = 1/2, the dual process $\tilde{\eta}_n$ is an annihilating random walk. Either (m,n) → (m+1,n+1), or (m,n) → (m−1,n+1), so we can think of the particle at m at time n flipping a coin to decide whether to jump to m+1 or m−1 at time n+1. Again, if two particles land on the same site they annihilate each other.

For other values of a and b, the dual exhibits a combination of branching and random walking, so the system is called an annihilating branching random walk. The continuous time analogues of these processes were studied by Bramson and Gray (1985). It is easy to see that if there is no branching then the system dies out, in the sense that starting from all sites occupied, the density of occupied sites goes to zero. Bramson and Gray (1985) showed that if the branching occurs at a much faster rate than the random walking then the process has positive probability of not dying out.

As in our study of the contact process, the duality equation can be used to study the asymptotic behavior of the process. The first step is to try to construct a stationary distribution. Let $\eta_n^{1/2}$ denote the system starting from product measure with density 1/2. We choose to start with product measure with density 1/2 because then the duality equation gives:

$$(2) \qquad P(\, |\eta_n^{1/2} \cap B| \text{ is odd }) = P(\, |\tilde{\eta}_0^{1/2} \cap \tilde{\eta}_n^B| \text{ is odd }) = (1/2) \cdot P(\, \tilde{\eta}_n^B \neq \phi \,).$$

ϕ is an absorbing state for $\tilde{\eta}_n$, so $\eta_n^{1/2}$ converges to a limit $\eta_\infty^{1/2}$ as n → ∞. The last observation shows that starting η_n from product measure with density 1/2 is like starting one of the systems in Section 5b (or 5c) from all sites occupied, and leads to the following conjecture. For the models discussed in this section,

(3) $$\eta_n^A \Rightarrow P(\ \tau^A < \infty\ )\ \delta_\phi + P(\ \tau^A = \infty\ )\ \eta_\infty^{1/2}$$

where $\tau^A = \inf\{\ n : \eta_n^A \neq \phi\ \}$. Bramson's result, mentioned above, allows us to prove that the last result is true for models of the form (p,0) with p near 1. To prove the result in general for these models, you will have to find a characterization of the critical value like the one introduced in Section 4a.

The connection between cancellative and additive models hinted at in the discussion of (2) is a little more than an analogy. Starting from the random graph at the beginning of the section, we can define two processes by

$$\xi_n^A(x) = (N_n^A(x) \wedge 1)$$
$$\eta_n^A(x) = N_n^A(x) \bmod 2$$

where $a \wedge 1 = \min\{a,1\}$. These processes have, in the notation of Section 5b,

$$\xi_n^A\ :\ f(1) = a+b \qquad f(2) = a+2b$$
$$\eta_n^A\ :\ f(1) = a+b \qquad f(2) = 2b$$

and they have duals defined by

$$\check{\xi}_n^A(x) = (\check{N}_n^A(x) \wedge 1)$$
$$\tilde{\eta}_n^A(x) = \check{N}_n^A(x) \bmod 2.$$

Oberving that $N \wedge 1 \geq N \bmod 2$, and using the two duality equations, gives

(4) $P(x \in \xi_n^{\mathbb{Z}}) = P(\xi_n^{\{x\}} \neq \phi) \geq P(\tilde{\eta}_n^{\{x\}} \neq \phi) = 2P(x \in \eta_n^{1/2}).$

So the density of $\eta_\infty^{1/2}$ is less than 1/2 than that of ξ_∞^1. This does not tell us much, but

in the case a = p and b = 0, it implies that the critical value of η is $\geq$ and probably $>$

than that of site percolation ($\approx$ .706). At the end of this section, we have simulations of

the process run for $0 \leq t \leq 360$ to support the last claim. The first picture shows the

process with p = .78, and it is fairly clear that the system is dying out. The second

picture shows the process at p = .82, the value of p_c predicted by Kinzel's finite size

scaling calculations. It looks to me like the process is surviving. In the last picture, p =

.86 and the process is doing even better. To close our discussion of p_c, we would like to

point out that when Vichniac, Tamayo, and Hartman (1986) reproduced Kinzel's

figure they erased the dashed line. These authors study the models with $p_2 = 1 - p_1$,

and they make a number of claims based on simulations. We invite the reader to look

at the paper and try to separate the wheat from the chaff.

Our discussion of cancellative systems would not be complete without some

mention of the model that sits at the top of the triangle: a = 0, b = 1/2. In this case, ξ

= η is the discrete time voter model, ξ is coalescing random walk, and $\tilde{\eta}$ is annihilating

random walk. For results that show that in continuous time $\eta_t^{\mathbb{Z}}$ is asymptotically a

1/2–thinning of $\xi_t^{\mathbb{Z}}$, see Arratia (1981), who gives the rates at which the densities of

these processes go to zero, and the limiting point processes that result when $\xi_t^{\mathbb{Z}}$ and $\eta_t^{\mathbb{Z}}$

are properly rescaled. In our study of the voter model in Chapter 2, we used the dual ξ.

For a use of the dual $\tilde{\eta}$ in the study of the voter model, see Presutti and Spohn (1983).

For more about annhilating and coalescing random walks, and cancellative duality, see

Griffeath (1979).

P = .78

P = .82

P = .86

6 Percolation in Two Dimensions

6a Bond Percolation

Percolation is one of the simplest systems that exhibits a "phase transition." In this model we consider $\mathbb{Z}^2$ as a graph with edges connecting each pair of points x,y with $|x-y| = 1$. The edges (which we will call bonds) are designated as open or closed with probabilities p and 1–p respectively, and the choices are made independently for each bond. We think of the open bonds as being air spaces that are large enough to permit the passage of a fluid. With this in mind we make the following definitions:

$x \to y$ (y can be reached from x) if there is an open path from x to y; that is, there is a sequence $x_0 = x, \ldots x_n = y$ of points in $\mathbb{Z}^2$ such that for each $m \leq n$ the bond from x_{m-1} to x_m is open.

The cluster containing 0, $C_0 = \{ x : 0 \to x \}$.

C_0 is the set of sites that will become wet if there is a source of fluid at 0. If C_0 is infinite, then we say percolation occurs. The probability of percolation is a nondecreasing function of p, so it is natural to define the critical probability

$$p_c = \inf \{ p : P (|C_0| = \infty) > 0 \}$$

The first thing to be proved is that p_c is nontrivial, that is, $0 < p_c < 1$. To prove the lower bound, we begin by observing that if there is a path from 0 to x then there is a self–avoiding path; that is, one that visits any point at most once. The number of self–avoiding paths of length n is less than $4(3)^{n-1}$. (The first step may be in any direction but after that we are forbidden to go back where we came from.) The probability that such a path is open is p^n, so the probability of having an open self–avoiding path of length $\geq$ N is at most

$$\sum_{n=N}^{\infty} 4(3)^{n-1} p^n.$$

If $p < 1/3$, then the last sum $\to 0$ as $N \to \infty$, so $P(\,|C_0| = \infty\,) = 0$ and $p_c \geq 1/3$.

To prove $p_c < 1$, we could observe that the critical value for (unoriented) percolation is less than that for the oriented case, which is $\leq 8/9$ by Section 5a. However, we have advertised that this chapter is independent of the previous ones and the argument introduces some notions that will be useful later, so we will show that $p_c < 1$ by modifying the proof we gave for the oriented percolation. Let A = { (0,0), ... (N,0) }, let C = { y : there is an x $\in$ A with x $\to$ y }, and let

$$W = \bigcup_{x \in C} x + [-1/2, 1/2]^2.$$

W stands for wet region. In words, we have replaced each point in C by a square of side 1 centered at the point, so the result is a solid blob. If C is finite, let Γ be the boundary of the unbounded component of the open set $R^2 - W$. Γ is called the contour associated with C.

If we let $Y^2 = (1/2, 1/2) + Z^2$, and consider Y^2 as a graph with edges connecting points x,y $\in Y^2$ with $|x-y| = 1$, then drawing a picture (see Figure 6.1), one sees that Γ is a union of edges on the graph Y^2, and each edge in Γ crosses a closed

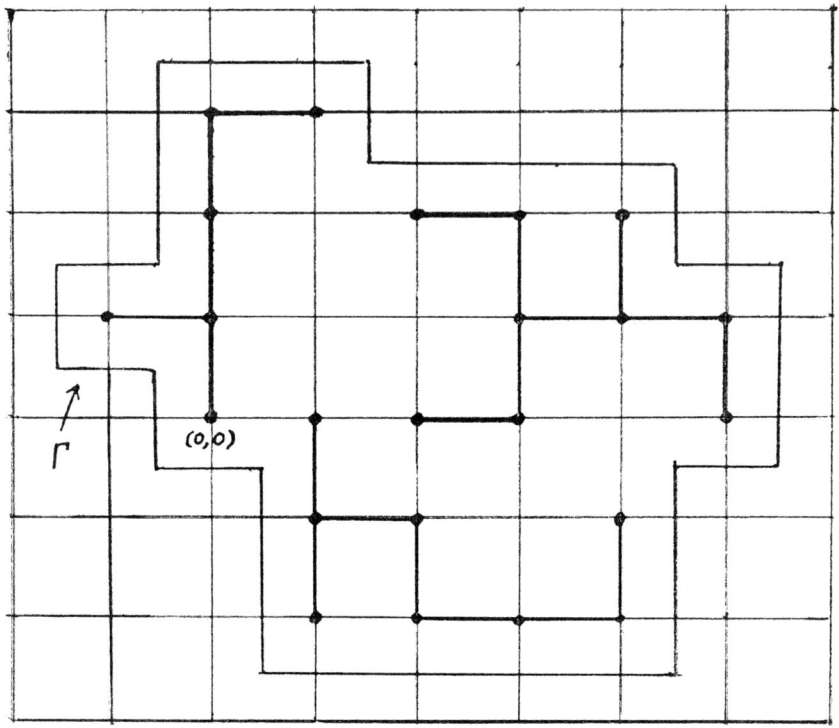

Figure 6.1

bond on $\mathbb{Z}^2$. The last sentence becomes easier to say if we declare the bonds on Y^2 to be open (resp. closed) if the bonds on $\mathbb{Z}^2$ which they cross are open (resp. closed), so we will use this convention in what follows. If the contour has length n (i.e., it consists of n edges), then the probability all the edges it contains are closed is $(1-p)^n$. The contour contains some edge of the form $((k+1/2,-1/2),(k+1/2,1/2))$ where k is a nonnegative integer $< n$, and it never traverses the same bond twice, so the number of contours of length n is less than $n\,3^{n-1}$. The shortest possible contour has length $2N+4$, so the probability that a contour exists is at most

$$\sum_{n=2N+4}^{\infty} n \, 3^{n-1}(1-p)^n.$$

If $p > 2/3$ and N is large, we have

$$0 < P \, (\, C \text{ is infinite }) \leq (N+1) \, P(\, C_0 \text{ is infinite }),$$

since C is infinite if and only if the cluster containing some point in A is.

The last two results show that we have $1/3 \leq p_c \leq 2/3$, and if you are good at leaping to conclusions, you can guess that the answer is the midpoint of this interval: $p_c = 1/2$. The first step in proving this was taken by Harris in 1960 when he showed $p_c \geq 1/2$. Four years later two physicists, Sykes and Essam, showed that if one assumed that the function

$$f(p) = \sum_{n=1}^{\infty} n^{-1} \, P_p(\, |C_0| = n \,)$$

(the subscript on P indicating that bonds are open with probability p) has a singularity at $p = p_c$ but is otherwise smooth, then $p_c = 1/2$. (f(p) gives "the number of clusters per unit volume;" that is, the limit of (the # of clusters in $[-n,n]^2)/(2n+1)^2$.)

This state of affairs existed for over 15 years, until Kesten showed in (1980a) that $p_c = 1/2$. The keys to Kesten's proof are the idea of a sponge crossing and a "duality" for planar percolation processes, hinted at in the proof of $p_c \leq 2/3$. A sponge is a rectangle $[0,m]x(0,n)$, and we say a left to right crossing occurs if there is a path of open bonds from $\{0\}x(0,n)$ to $\{m\}x(0,n)$. A little graph theory (due to Whitney in the 30's) shows that either there is a left to right crossing of $[0,m]x(0,n)$ on Z^2, or there is a closed path on the dual graph Y^2 (described above) from the top to bottom of

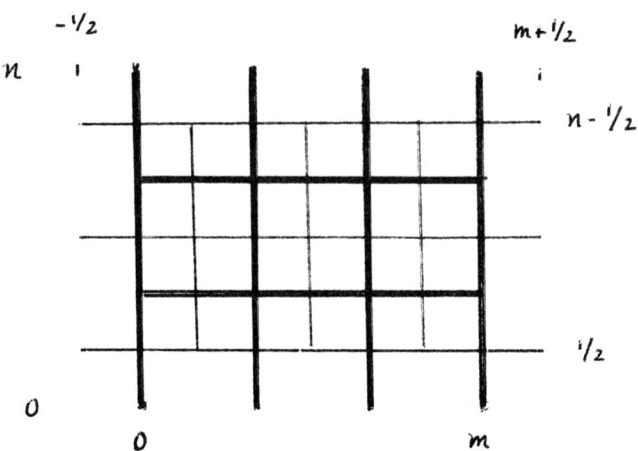

Figure 6.2

$(-1/2, m+1/2) \times [-1/2, n+1/2]$, but not both. See Figure 6.2 for a picture. For a proof of this and the other facts from graph theory we will use below, see Chapter 2 of Kesten (1982). We will prove something for oriented percolation that implies this in Chapter 9.

When $n = m + 1$, the original sponge and the dual sponge are both n−1 units long and n units wide, so if $p = 1/2$ it follows from symmetry that the probability of a left to right crossing is $1/2$. To get the last conclusion, and have a perfect correspondence between the bonds in the original and the dual sponge, we have made one interval closed and the other one open. While this convention makes the last statement neat, it will make life harder later, so we would like to observe that the bonds on $\{0\} \times (0,n)$ and $\{n\} \times (0,n)$ are not useful for finding a left to right crossing, so we can and will make our sponges products of open intervals in what follows. (We apologize to the reader for dwelling on such minutia. Having the right definition of sponges will be important at several places below.)

The observation that the probability of crossing $(0,n{-}1) \times (0,n)$ is $1/2$ when $p = 1/2$ is one of the keys to the proof that $p_c \geq 1/2$. The second is the following useful result which is usually referred to as

Harris' inequality. Let $X_1,\ldots X_n$ be independent random variables, and let f and g be bounded nondecreasing functions on $\mathbb{R}^n$; that is, if $x_i \leq y_i$ for all i then $f(x) \leq f(y)$. If we let $X = (X_1, \ldots X_n)$ then

$$E(f(X)g(X)) \geq Ef(X)\, Eg(X).$$

Proof: We prove the result by induction on n. If n=1 then for all x and y we have

$$(f(x){-}f(y)) \cdot (g(x){-}g(y)) \geq 0.$$

(Consider two cases $x \geq y$ and $x \leq y$.) If we replace x by X, and y by X', where X' is independent of X and has the same distribution, then we get

$$0 \leq E[(f(X){-}f(X')) \cdot (g(X){-}g(X'))] = 2\, E[f(X)g(X)] - 2\, Ef(X) \cdot Eg(X),$$

the result for $n = 1$.

To prove the result in general, observe that applying the result for $n = 1$ to $f_1(x) = f(x,x_2,\ldots x_n)$ and $g_1(x) = g(x,x_2,\ldots x_n)$ gives

$$E(f(X)g(X)\,|\,X_2,\ldots X_n) \geq E(f(X)\,|\,X_2,\ldots X_n)\, E(g(X)\,|\,X_2,\ldots X_n).$$

Taking expected values, and applying the result to $f_2(x_2,\ldots x_n) = Ef(X_1,x_2,\ldots x_n)$,

$g_2(x_2,...x_n) = Eg(X_1,x_2,...x_n)$, which are functions of n–1 variables, now gives the desired result.

In this section, we are mostly interested in the following special case which we also refer to as Harris' inequality.

(1) **Lemma.** Let A and B be increasing events; that is, if A occurs for some configuration of open and closed bonds, then it does in every outcome with more open bonds. Then

$$P(\ A \cap B\ ) \geq P(A)P(B)$$

Proof: If A and B depend upon only finitely many bonds, then this is a consequence of the last result, since the indicator functions 1_A and 1_B are bounded nondecreasing functions. To extend this to general A and B, list the bonds in Z^2 in some order and let $\mathscr{F}_n$ be the σ–field generated by bonds 1,...n. If we let $A_n = \{\ P(A\,|\,\mathscr{F}_n) \geq 1/2\ \}$ then A_n is an increasing event which only depends upon the state of bonds 1,... n. The martingale convergence theorem implies that $1_{A(n)} \to 1_A$ a.s. as n → ∞, so $P(A_n \vartriangle A) \to 0$ as n → ∞. If we define B_n in the analogous way, apply Harris' inequality to A_n and B_n, and let n → ∞, we get (1).

The existence of an open path from one point (or set of points) to another is the basic example of an increasing event. If A and B are both events of this type, then the lemma says that if we are lucky enough to find one connection, then it increases the probability of finding the other one. The third and final ingredient in our proof is the following weird and wonderful result from Russo (1981).

(2) Lemma. Let $\rho_{m,n}$ be the probability of a left to right crossing of an mxn sponge. Then

$$\rho_{3L/2,L} \geq (1-(1-\rho_{L,L})^{1/2})^3.$$

Pep talk: As you can probably guess from looking at the right–hand side, the proof is a little tricky. However, generalizations of (2) will be important for developments in Chapters 7 and 9, so we must prove it in detail. This does not mean you have to read the proof right now. Once (2) is proved, the rest of the proof is relatively painless, so if you get bored or confused by the details you can safely skip ahead to (3) below.

Proof of (2): The first step explains the unusual formula in the answer.

The square root trick. If A_1 and A_2 are two increasing events with $A = A_1 \cup A_2$ and $P(A_1) = P(A_2)$, then

$$P(A_1) \geq 1 - (1-P(A))^{1/2}.$$

Proof: Using $P(A_1) = P(A_2)$, (1), and, $A = A_1 \cup A_2$ gives

$$\begin{aligned}
(1-P(A_1))^2 &= 1 - P(A_1) - P(A_2) + P(A_1)P(A_2) \\
&\leq 1 - P(A_1) - P(A_2) + P(A_1 \cap A_2) \\
&= 1 - P(A).
\end{aligned}$$

Rearranging gives the desired result.

Remark: In Chapters 7 and 9 we will need to generalize (2) to models in which the bonds are not independent, but (1) holds. To prepare for our later claims, you should observe that the proof of the square root trick only uses (1) and set theory, so the parts of the argument that only use (1) and (2) will generalize immediately. When all is said and done, this leaves only two places where independence is used and the argument has to be modified.

The nice thing about the last inequality is that if $P(A)$ is close to 1, then so is the lower bound on $P(A_1)$, and that is much better than just saying $P(A_1) \geq P(A)/2$. In proving (2), we apply the last result to get crossings that begin or end in the top (or bottom) half of certain squares. If one does this cleverly, left to right crossings of $(0,L)\times(0,L)$ and $(L/2,3L,2)\times(0,L)$ and a top to bottom crossing of the second square can be combined to get a crossing of $(0,3L/2)\times(0,L)$. Carrying this out requires making a large number of definitions. The reader should refer to Figure 6.3 as we go along.

Let E_s be the event that s is the lowest left to right crossing of $(0,L) \times (0,L)$.

Let s_r be the portion of this path from the time it last hits $\{L/2\} \times (0,L)$ until it reaches $\{L\} \times (0,L)$ (the thick line in Figure 6.3).

Let s_{rr} = the reflection of s_r through $\{L\} \times (0,L)$ (the dotted line in Figure 6.3).

Let $\mathscr{A}(s_r \cup s_{rr})$ = the points in $(L/2, 3L/2) \times (0,L)$ strictly above $s_r \cup s_{rr}$.

Let $\mathscr{B}(s)$ = the points in $(0,L) \times (0,L)$ strictly below s (the shaded region in Figure 6.3).

Let F_s be the event that there is a path starting from $(L/2,3L/2) \times \{L\}$ and connected to s_r in $\mathscr{A}(s_r \cup s_{rr}) - \mathscr{B}(s)$; notice that this definition allows you to use s, which

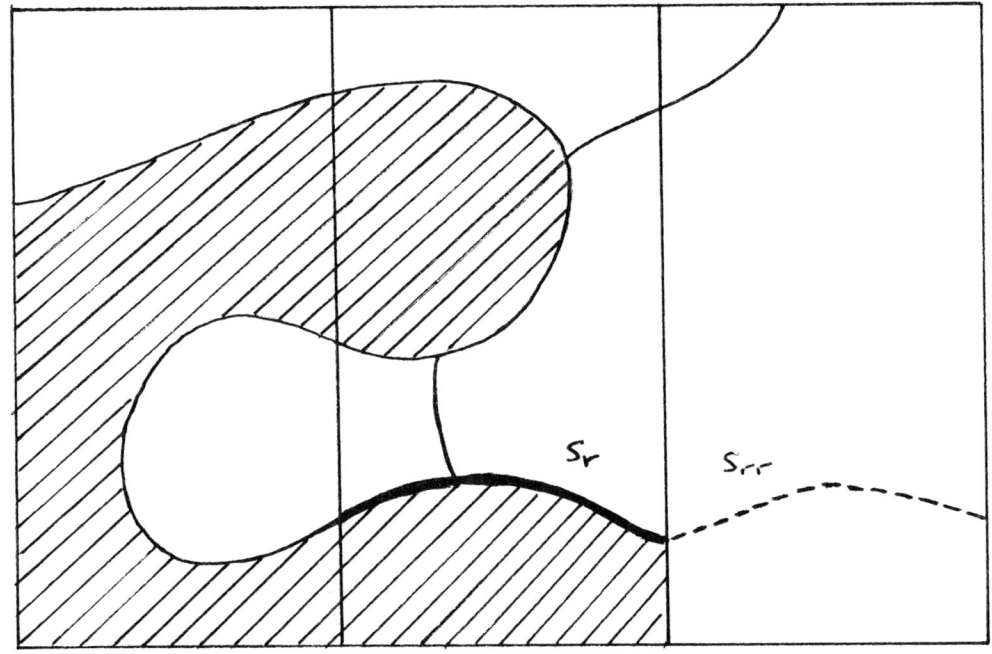

Figure 6.3

consists of open bonds.

Let G be the union of $E_s \cap F_s$ over all the paths s for which the first point of s_r has y coordinate $\leq L/2$ (like the one drawn in Figure 6.3).

Let H be the event that there is a left to right crossing of $(L/2, 3L/2) \times (0, L)$ which starts at a point with y coordinate $\geq L/2$.

We have not drawn a path of the last type in Figure 6.3, but we invite you to do so now to convince yourself that on $G \cap H$ there is a left to right crossing of $(0, 3L/2) \times (0, L)$. To prove (2) then, it suffices to show

$$P(G \cap H) \geq (1 - (1 - \rho_{L,L})^{1/2})^3.$$

The first step in doing this is to observe that Harris' inequality implies $P(G \cap H) \geq$ $P(G)P(H)$, and using the square root trick with $A = \{$ there is a crossing of $(L/2, 3L/2)$ $\times (0,L)\}$, and $A_1 = H$ gives

$$P(H) \geq (1 - (1 - \rho_{L,L})^{1/2}).$$

To estimate $P(G)$ we write

$$P(G) = \sum_s P(E_s \cap F_s) = \sum_s P(E_s)P(F_s | E_s),$$

and observe that if F_s' is the event that there is a path from $(L/2, 3L/2) \times \{L\}$ to s_r in $\mathscr{A}(s_r \cup s_{rr})$ then

(*) $P(F_s | E_s) = P(F_s) \geq P(F_s').$

Here we use two things: (i) E_s is measurable with respect to the sites in $\overline{\mathscr{B}}(s) = \mathscr{B}(s)$ $\cup s$; (ii) the presence of s which is open makes it easier to find the connections we want. Notice that in (i) we use independence, and this is the first time we have used something more than Harris' inequality.

   With (*) established, the rest is easy. Two more applications of the square root trick gives

$$P(F_s') \geq (1 - (1 - \rho_{L,L})^{1/2}),$$

and

$$\sum_{s} P(E_s) \geq (1 - (1-\rho_{L,L})^{1/2}).$$

Putting the pieces together we have (2).

With (2) in hand, the rest of the proof of $p_c \geq 1/2$ is smooth sailing. Drawing a picture (see Figure 6.4) shows

(3) $$\rho_{kL,L} \geq (\rho_{(k+1)L/2,L})^3 \text{ for } k \geq 1.$$

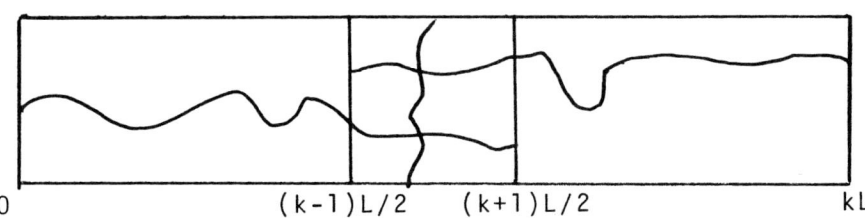

0   (k-1)L/2   (k+1)L/2   kL

Figure 6.4

For if all three paths exist, the desired crossing occurs, and the three events are increasing, so the probability of the intersection is bigger than the product of their probabilities. Using the last result with $k = 2$ and $k = 3$, gives

$$\rho_{2L,L} \geq (\rho_{3L/2,L})^3,$$

and

$$\rho_{3L,L} \geq (\rho_{2L,L})^3.$$

Now when $p = 1/2$, we have $\rho_{L,L} \geq \delta > 0$ for all L. (Recall $\rho_{L-1,L} = 1/2$ for all L, so $\rho_{L,L} \geq 1/4$. Later we will want to claim that the argument below goes through under the assumption $\rho_{L,L} \geq \delta > 0$, so we will not use δ instead of 1/4 below.) Combining the last observation with (2) and the last two results shows that when p = 1/2 we have

(4) $\rho_{3L,L} \geq (1 - (1-\delta)^{1/2})^{27}$.

The last lower bound is not very big, but has the virtue of being independent of L.

When p = 1/2, the last bound applies equally well to crossing dual rectangles with closed paths. If we can cross (n,3n)x(−3n,3n) and (−3n,−n)x(−3n,3n) from top to bottom, and cross (−3n,3n)x(n,3n) and (−3n,3n)x(−3n,−n) from left to right with closed paths on the dual then we create a closed circuit of dual bonds that forces C_0 to be finite. The probability all four paths exist is, by (4) and Harris' inequality, at least

$$(1 - (1-\delta)^{1/2})^{108} > 0.$$

The last probability is ridiculously small, but taking n = 3^k gives an infinite sequence of disjoint square annuli in which the event can happen. So with probability 1 it will eventually happen and C_0 will be finite.

Having shown that $p_c \geq 1/2$, the next step is to show that if p > 1/2 then percolation occurs with positive probability. This is a two step process:

(i) first show there is an $\epsilon_0 > 0$ so that if $\rho_{L,L} > 1 - \epsilon_0$ then percolation has positive probability, and then

(ii) show that if p > 1/2 then $\rho_{L,L} \to 1$ as L → ∞.

Given the developments above, the first step is easy. We start by observing that (3) could have been written as

(5) $$1- \rho_{kL,L} \leq 3(1 - \rho_{(k+1)L/2,L}),$$

since the probability that some path in Figure 6.4 fails to exist is smaller than the right—hand side. A similar argument (see Figure 6.5) shows

(6) $$1- \rho_{4L,L} \leq 5(1-\rho_{2L,L})$$

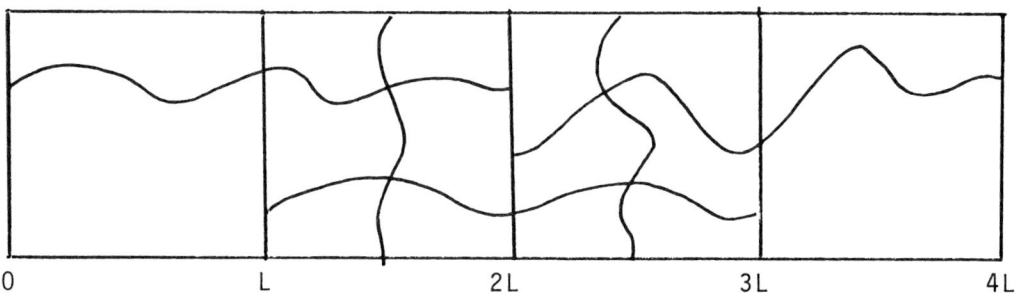

Figure 6.5

The second jab of our one—two punch is due to Aizenman, Chayes, Chayes, Fröhlich, and Russo (1983):

(7) $$\rho_{4L,2L} \geq 1 - (1-\rho_{4L,L})^{2}.$$

(7) is proved by placing two 4L by L rectangles next to each other, and observing that we will cross the resulting 4L by 2L rectangle unless both 4L by L rectangles fail to be crossed, and those events are INDEPENDENT. (Excuse me for shouting. Here we use independence for the second and final time. The reader should note that we need to use open sponges for the two events to be independent.)

Combining (6) and (7), gives

(8) $$\rho_{4L,2L} \geq 1 - 25(1 - \rho_{2L,L})^2.$$

At first (8) may not look like much, but if we start with a small value, squaring is much more powerful than multiplying by 25. To be less mystical, observe that if $\rho_{2L,L}$ $= 1 - \lambda/25$ with $\lambda < 1$ then (8) implies

$$\rho_{4L,2L} \geq 1 - \lambda^2/25,$$

$$\rho_{8L,4L} \geq 1 - \lambda^4/25,$$

and by induction,

(9) $$\rho(2^k L, 2^{k-1} L) \geq 1 - (1/25)\exp(2^{k-1} \log \lambda).$$

The last inequality shows that if $\rho_{2L,L}$ is close enough to 1, then the sponge crossing probabilities converge to 1 exponentially fast.

To get the last scheme started, we observe that if $\rho_{L,L} > 1 - \epsilon$ then (2) implies

$$\rho_{3L/2,L} \geq (1 - \epsilon^{1/2})^3,$$

and (5) gives us

$$\rho_{2L,L} \geq 1 - 3 (1 - \rho_{3L/2,L}).$$

So if $\epsilon < \epsilon_0$ (10^{-6} is a possible choice), then $\rho_{2L,L} > .99$, and (9) holds with $\lambda = 1/e$. That is, we have

(10) The rescaling lemma. There is an $\epsilon_0 > 0$ so that if $\rho_{L,L} > 1 - \epsilon_0$ then

$$\rho(2^k L, 2^{k-1} L) \geq 1 - (1/25)\exp(-2^{k-1}).$$

To get from (10) to a positive probability of percolation, pick L so that $\rho_{L,L} > 1 - \epsilon_0$, define boxes

$$B_{2k-1} = (0, 2^{2k-1}L) \times (0, 2^{2k-2}L),$$
$$B_{2k} = (0, 2^{2k-1}L) \times (0, 2^{2k}L),$$

and observe that if we get left to right crossings of all the B_{2k-1} and top to bottom crossings of all the B_{2k}, then there is an infinite path starting on $\{0\} \times (0,L)$. (See Figure 6.6.) The probability we get all the paths we want, and in addition that all the bonds on $\{0\} \times [0,L]$ are open is, by Harris' inequality, at least

$$p^L \cdot \prod_{k=1}^{\infty} [1 - (1/25)\exp(-2^{k-1})] > 0,$$

and we have demonstrated (i).

Last but not least we come to (ii). The key to the proof of this is:

(11) Russo's formula. If A is an increasing event then

$$\frac{d}{dp} P_p(A) = E(\ \# \text{ of pivotal bonds }),$$

where a bond is said to be pivotal if changing its state changes the occurrence of the event; that is, when the bond is open A occurs but if it is closed it does not.

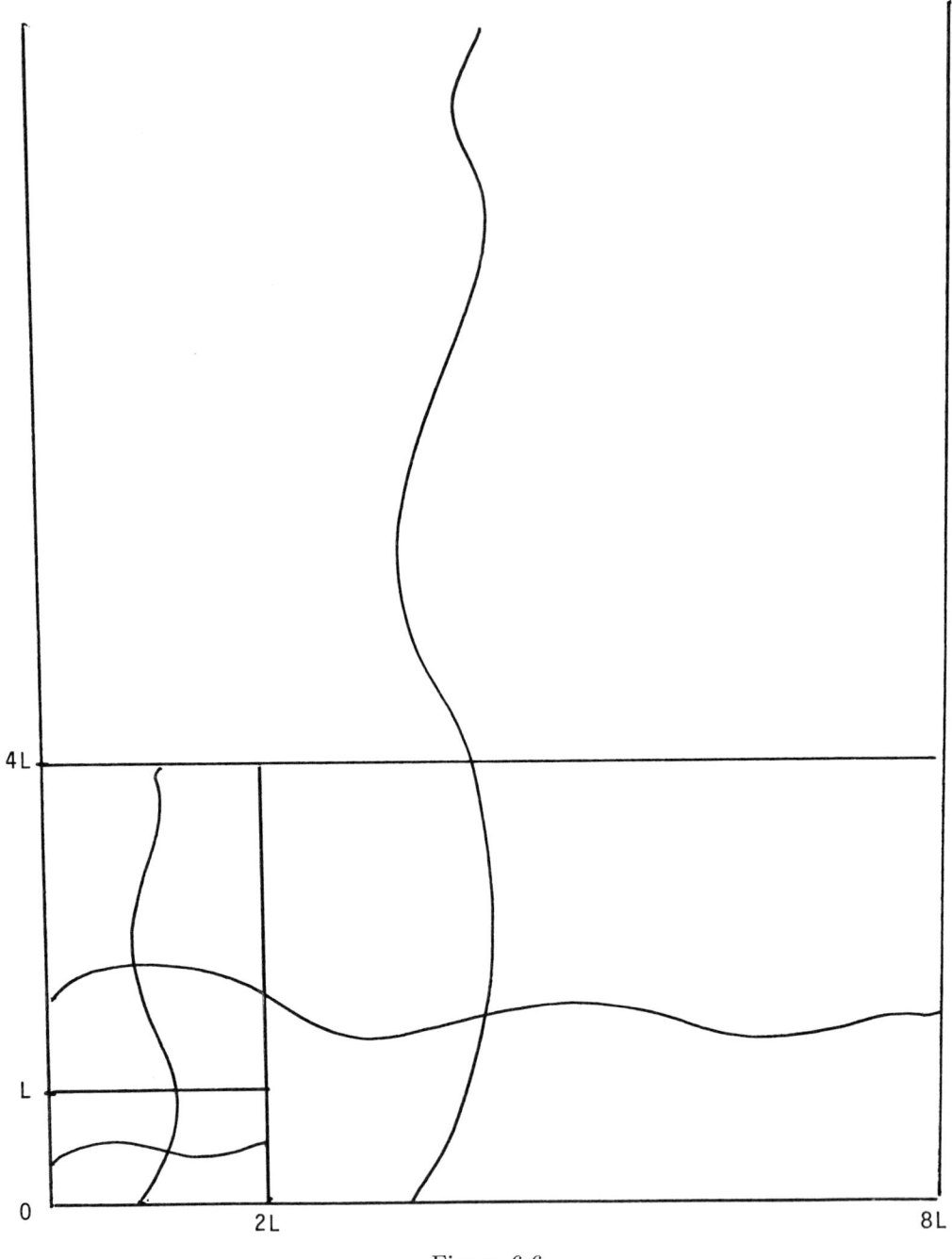

Figure 6.6

The proof of (11) is left as an exercise for the reader. (Start with the case of a finite graph. See Kesten (1982), p.77–80 for a solution.) To prove (ii) we apply (11) to $A_L =$ {there is a left to right crossing of $(0,L) \times (0,L)$ with L large}, and show that if $p \geq 1/2$, and $P_p(A_L) < 1 - \epsilon_0$ (the magic constant in (10)), then the derivative is large. Since $P_{1/2}(A_L) \geq \delta > 0$, the last fact implies that

$$\inf \{ p : P_p(A_L) > 1 - \epsilon_0 \} \to 1/2,$$

and in view of (10), it follows that $\rho_{L,L} \to 1$ for all $p > 1/2$.

To complete the proof now, we need to get our hands dirty and find a large number of pivotal bonds. The fact that this step is nontrivial can be established by noting that Russo (1978) knew that large sponge crossing probabilities implied percolation (and was aware of Russo's formula!), but he could not complete the proof, nor could anyone one else, until Kesten came along and finished the problem.

Kesten's idea was to look at the lowest left to right crossing. The event that σ is the lowest crossing is determined by the bonds that lie on or below σ. So if we condition on σ being the lowest crossing, the distribution of the bonds that lie above it is not effected by the conditioning. In the virgin territory that lies above σ, we look for a closed path on the dual which comes down and touches σ. If $P_p(A_L) \leq 1 - \epsilon_0$ then the probability of finding one is at least ϵ_0. (The probability of a top to bottom dual crossing by closed bonds is $1 - P(A_L)$, and all we want to do is go down to σ.) If there is a path on the dual from the top down to σ, pick the left most one and call it ω.

The bond at the point where ω touches σ is pivotal, since σ is the lowest crossing, and the presence of ω implies that every crossing must pass through this bond. In the region to the right of ω and above σ (which is still virgin territory), we look for more closed crossings by introducing roughly $\log_3 L$ disjoint square annuli

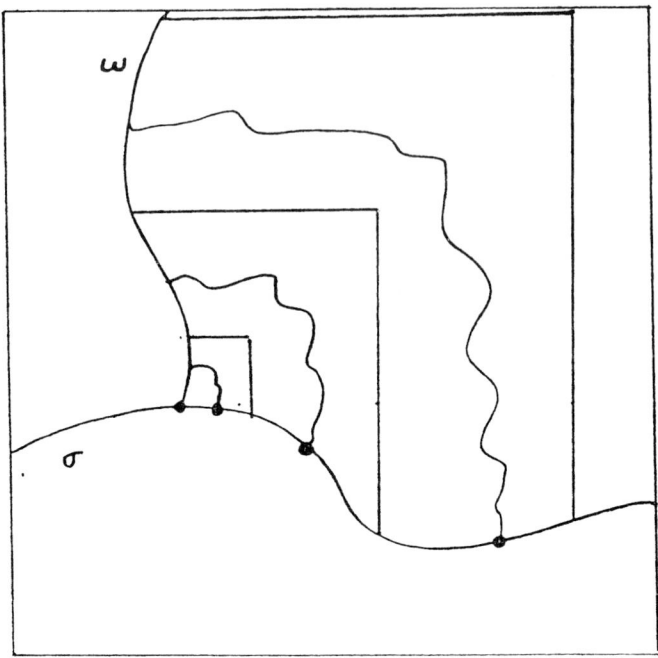

Figure 6.7

centered at the point where ω touches σ, and looking for closed circuits in these annuli that go from ω to σ. (See Figure 6.7.) Every circuit produces at least one pivotal bond, and when $P(A_L) \leq 1-\epsilon_0$ there is a lower bound on the probability of these circuits, so the expected number is at least $\delta \log L$ for some $\delta > 0$. For more details see Kesten (1980a) or (1982).

At this point we have shown that if $p > 1/2$ then C_0 is infinite with positive probability. For some results in Chapters 8 and 9 it will be useful to know

(12) Theorem. If $p > 1/2$, there is with probability one a unique infinite connected set of open bonds.

Proof: To prove this, observe that by the result for $p = 1/2$ there are infinitely many open circuits around 0. Because of this, the existence of an infinite component is measurable with respect to the state of bonds in $\mathbb{R}^2 - [-N,N]^2$ for any N. When $p > 1/2$, $P(\, C_0$ is infinite$) > 0$, so it follows from the 0–1 law* that $P(\,$ there is an infinite component$) = 1$. To finish the proof, if $x,y \in \mathbb{Z}^2$, then the first observation implies there is an open circuit with x and y in its interior, so if the clusters C_x and C_y containing x and y are infinite, then $C_x = C_y$.

*Note: Even though the random variables we are dealing with are indexed by the bonds in $\mathbb{Z}^2$, nothing fancy is being used here, just the ordinary 0–1 law for sequences. To apply it here, we list the bonds in $\mathbb{Z}^2$ in order by sorting them first by the distance from their midpoint to the origin, and then imposing some arbitrary order within each class (e.g., list them in clockwise order starting with the one on the x–axis). If we do this then the event is measurable with respect to the tail σ–field of this sequence.

6b Site Percolation

In this section we consider the percolation model in which sites in $\mathbb{Z}^2$ are independently designated as open and closed with probabilities p and 1–p respectively. This model can be analyzed by using the techniques we applied to the bond percolation in the last section, but a few new difficulties arise because the model is not "self–dual." (This phrase will be explained in a minute.) Since we will need to use results for site percolation in Chapter 7, we will now indicate how things have to be modified for the new model. We begin with the basic definition:

$x \to y$ (y can be reached from x) if there is an open path from x to y; that is, there is a sequence $x_0 = x, ... x_n = y$ of open sites so that for each $m \le n$, x_{m-1} and x_m are neighbors (i.e., $|x_{m-1} - x_m| = 1$).

Once $x \to y$ is defined, we can define the cluster containing 0, $C_0 = \{ x : 0 \to x \}$, and p_c as before. Again, the first thing to be proved is that $0 < p_c < 1$. Half of this is a consequence of the result in the last section:

(1) Theorem. $p_c(\text{site}) \ge p_c(\text{bond}) = 1/2$.

Proof: This is true for a general graph (see Hammersley (1961) or McDiarmid (1980), Theorem 4.1). We will prove the result in our special case by using the "bond to site transformation," which converts bond percolation problems into site percolation problems. Let $S = \{ (x+y)/2 : x,y \in \mathbb{Z}^2 \ |x-y| = 1 \}$. In words, the points in S are midpoints of the bonds in the graph we considered in the last section. We make S into a graph by connecting points u,v in S if the corresponding bonds have an endpoint in common. Site percolation on S is equivalent to bond percolation on $\mathbb{Z}^2$ if we consider clusters in the bond model to be sets of bonds. The graph S, however, can be obtained from $\mathbb{Z}^2$ by connecting diagonally adjacent vertices in every other square, rotating 45 degrees, and scaling the result by dividing by $\sqrt{2}$. (See Figure 6.8.) Since S has more edges than Z^2, site percolation on S (= bond percolation on Z^2) must have a smaller critical value.

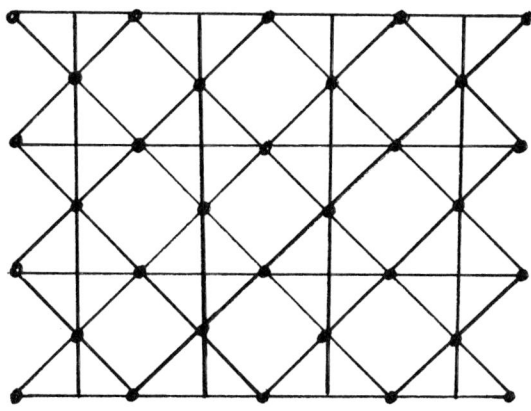

Figure 6.8

Exercise: Can you improve the last result to $p_c(\text{site}) > p_c(\text{bond})$? This is true but rather difficult to prove. For a solution, see Kesten (1982), Chapter 10.

(1) shows that $p_c(\text{site}) \geq 1/2$. To show that $p_c < 1$ (here and in what follows we will drop the "(site)" when it is clear we are talking about site percolation), we again use a contour argument but the contours are different. Let $C_0 = \{ x : 0 \to x \}$, and if C_0 is finite, let $B_0 = \{ y : y$ is adjacent to a point in C_0, and y can be connected to infinity in C_0^c; that is, there is an infinite self–avoiding path starting at y that lies in $C_0^c\}$. Here adjacent means $|x-y| = 1$, and a self–avoiding path is a sequence of

adjacent sites all of which are different. B_0 is the "external boundary of C_0." See
Figure 6.9 for a picture. The c's indicate points in C_0. The b's are points in B_0. The
x's are closed sites which are not part of B_0.

```
            b  b  b  b
         b  c  c  c  c  b
         b  c  x  c  c  b
         b  c  c  c  b  b
      b  c  c  x  x  c  b
         b  c  c  c  c  b
         b  b  b  b
```

Figure 6.9

B_0 is not connected as a subset of the graph $(Z^2, \mathscr{E})$, where $\mathscr{E}$ is the usual set of
edges $\{ (x,y) : |x-y| = 1 \}$, but it is connected as a subset of the graph $X^2 = (Z^2, \mathscr{E}^*)$
where $\mathscr{E}^* = \{ (x,y) : x \neq y$ but $|x_1-y_1|$ and $|x_2-y_2| \leq 1 \}$, that is, in addition to the
usual edges there are edges to diagonally adjacent points: $z \to z + (1,1)$, $z \to z + (1,-1)$,
$z \to z + (-1,1)$, and $z \to z + (1,-1)$. We call X^2 the dual graph because, like Y^2 in the
last section, C_0 is finite if and only if there is a closed circuit on the dual that
surrounds 0. (For a proof of this, and the other facts from graph theory that we will
use below, see Chapter 2 of Kesten (1982).) With the last fact in hand, it is easy to
repeat the contour argument in the last section to conclude

(2) Theorem. $p_c < 6/7$.

The bound comes from the fact that there are $\leq n(7)^{n-1}$ contours of length n. The proof is left as an exercise for the reader.

As in the last section, the graph duality introduced in the contour argument above is one of the three key ingredients. The other two are:

(3) Harris' inequality. Let A and B be increasing events; that is, if A occurs for some configuration of open and closed sites then it does in every outcome with more open sites. Then

$$P(A \cap B) \geq P(A)P(B).$$

(4) Russo's lemma. Let $\rho_{m,n}$ be the probability of a left to right crossing of an m x n sponge. Then

$$\rho_{3L/2,L} \geq (1 - (1 - \rho_{L,L})^{1/2})^3.$$

The proof of (3) for site percolation is the same as for bond percolation, since it comes from a general fact about increasing functions of independent random variables. The proof of (4) is also identical. The second claim is harder to believe, since it involves checking that the tricky proof given in the last section works for site percolation. The reader should instantly lose her skepticism when she hears that the proof in Russo's paper is for site percolation, and what we have done in the last section is rewrite his proof for bond percolation.

While the last two results are the same as (1) and (2) in the last section, the treatment of site percolation diverges at the next step, which in the case of bond percolation was to show $p_c \geq 1/2$ by observing that the probability of crossing an (n+1) x n sponge was 1/2 when p = 1/2. For site percolation in $\mathbb{Z}^2$, the dual is site percolation in X^2, so we cannot use symmetry to locate p_c and an indirect approach is needed. Skipping over the proof of $p_c \leq 1/2$ in the last section, which does not generalize, we find that the arguments for $(5) - (9)$ can be repeated for site percolation to prove:

(5) The rescaling lemma. There is an ϵ_0 so that if $\rho_{L,L} > 1 - \epsilon_0$, then

$$\rho(2^k L, 2^{k-1} L) \geq 1 - (1/25)\exp(-2^{k-1})$$

for all $k \geq 1$, and consequently, there is positive probability of percolation.

The magic constant ϵ_0 is the same, because the constants that appear in the other inequalities are unchanged.

Applying the last reasoning to the dual graph X^2, and using the fact that either there is a left to right crossing by open sites, or a top to bottom crossing on the dual by closed sites gives

(6) Lemma. If $\rho_{L,L} < \epsilon_0$ then

$$\rho(2^{k-1} L, 2^k L) \leq (1/25) \exp(-2^{k-1}),$$

and consequently, the probability of percolation is 0.

To obtain the second conclusion, observe that if 0 is connected to a point on the boundary of $[-2^k, 2^k]^2$ then there is a crossing of $(0, 2^k)$ x $(-2^k, 2^k)$ or of one of the other three sponges obtained by rotating this one by 90, 180, or 270 degrees. (If 0 is connected to $\{2^k\}$ x $(-2^k,2^k)$ inside $[-2^k,2^k]^2$, then follow the path backward from the end until it hits $\{0\}$ x $(-2^k,2^k)$. See Figure 6.10. The path from a to b is the desired crossing.)

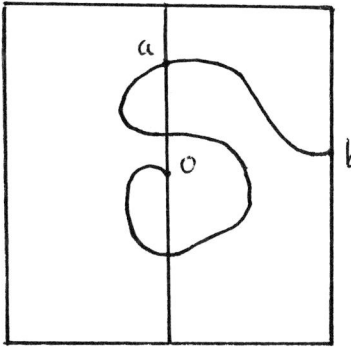

Figure 6.10

With (5) and (6) in hand, the denouement is the same as the proof of (ii) in the last section. We use Russo's formula ((11) In Section 6a) to conclude:

(7) $$\frac{d}{dp}\rho_{L,L} = E(\# \text{ of pivotal sites})$$

By an explicit construction, we find enough pivotal sites to conclude that if $\rho_{L,L} \in [\epsilon_0, 1-\epsilon_0]$, the right–hand side is $\geq c_L$, a sequence of constants that $\to \infty$ as $L \to \infty$. The last conclusion, when combined with (5) and (6), implies that the length of the interval $\{\, p : \rho_{L,L}(p) \in [\epsilon_0, 1-\epsilon_0] \,\}$ converges to 0 as $L \to \infty$. For results below, it will be useful to know that p_c is always in the interval. To prove this, we observe that if $\rho_{L,L}(p_c) > 1 - \epsilon_0$, then for some $p < p_c$ we would have $\rho_{L,L}(p) > 1-\epsilon_0$, and with (5), this contradicts the definition of p_c. On the other hand, if $\rho_{L,L}(p_c) < \epsilon_0$, then for some $p > p_c$ we would have $\rho_{L,L}(p) < \epsilon_0$, and with (6), this contradicts the definition of p_c.

The arguments above give us the following results:

(8) Theorem. If $p < p_c$ there are contants C, $\gamma \in (0,\infty)$ so that

$$P(\text{ radius of } C_0 > k) \leq C\, e^{-\gamma k}.$$

Proof: Since $\rho_{L,L}(p) \to 0$ for $p < p_c$, this follows from (6) above.

Remark. By working harder one can show that $P(|C_0| > k) \leq C \cdot e^{-\gamma k}$. See Section 5.1 of Kesten (1982). We will introduce the construction which is the key to the proof when we prove (2) in Section 8b.

(9) Theorem. When $p = p_c$ the probability of percolation is 0.

Proof: The probability of crossing an L x L sponge does not go to 1 at p_c, and hence the probability of crossing an L x L sponge by a closed dual path does not go to 0. Repeating the proof of $p_c \geq 1/2$ from the last section now shows there is no percolation at p_c.

(10) Theorem. If p_c^* is the critical value for site percolation on X^2 then $p_c + p_c^* = 1$ and there is no percolation when $p = p_c^*$.

Proof: Since $\rho_{L,L}(p) \to 0$ for $p < p_c$, the proof of (6) implies that if $p < p_c$ then vacant sites (which have density 1–p) percolate on X^2, and hence $p_c^* \leq 1-p_c$. Since the probability of an open crossing on an L x L sponge does not go to 0 at p_c, it follows from the proof of (9) that vacant sites do not percolate on X^2 when $p = p_c$ and hence

$$p_c^* \geq 1 - p_c.$$

The results above tell us quite a bit about what site percolation looks like for $p < p_c$, $p = p_c$, and $p > p_c$, but do not tell us much about the value of p_c: $1/2 < p_c \leq 6/7$. To get an idea of what p_c is, we use the fact that

$$p_c = \sup \{ \, p : \rho_{L,L}(p) \to 0 \, \} = \inf \{ \, p : \rho_{L,L}(p) \to 1 \, \},$$

and resort to computer simulation. The four pictures show the set of sites that can be reached from the left edge when $L = 200$, and $p = 146/256$, $150/256$, $154/256$, and $158/256$, respectively. (To see the reason for our denominators, recall $2^8 = 256$ and a byte is 8 bits.) The second fraction is .5859 and the third is .6016, so the pictures are consistent with the numerical result $p_c \approx .5923$. (See Djordevic, Stanley, and Margolina (1982).)

$$P = 146/256$$

P = 150/256

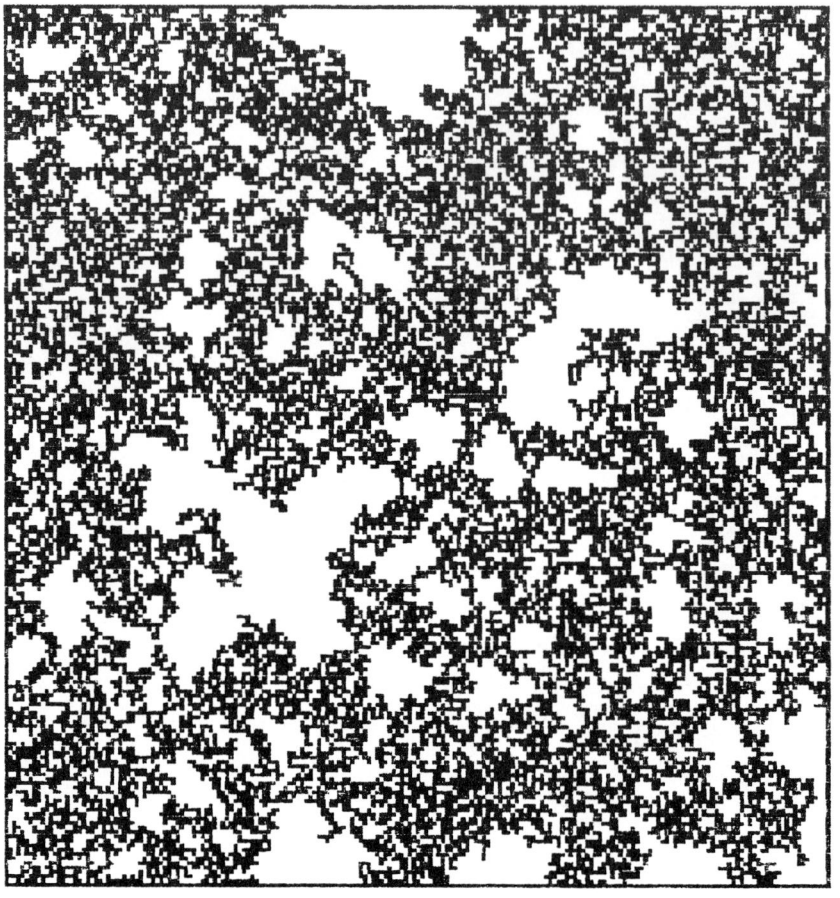

$$P = 154/256$$

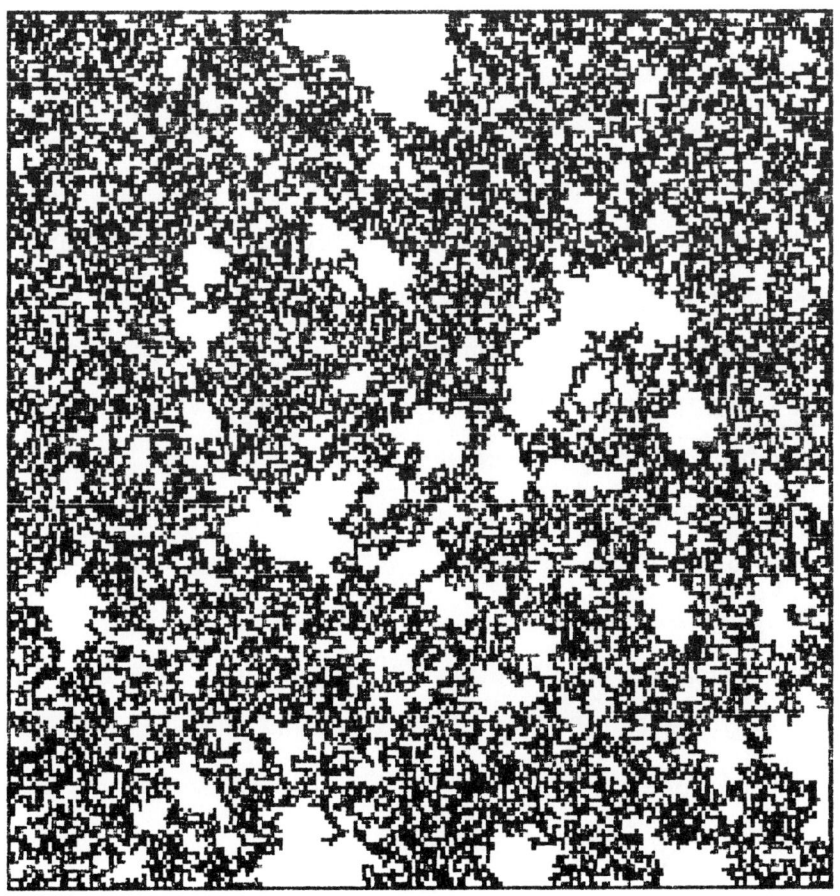

$$P = 158/256$$

7 Mandelbrot's Percolation Process

In (1974), Mandelbrot introduced a process in $[0,1]^2$ which he called "canonical curdling," and later used in his book(s) on fractals to generate self–similar random sets with Hausdorff dimension $D \in (0,2)$. In this chapter, we study the connectivity or "percolation" properties of these sets, proving all of the claims he made in Section 23 of the Fractal Geometry of Nature (1983), and a new one that he did not anticipate. There is a probability $p_c \in (0,1)$, so that if $p < p_c$ then the set is "dustlike"; that is, the largest connected component is a point, whereas if $p \geq p_c$ (notice the $=$) opposite sides of the square are connected with positive probability. More succinctly, the system has a "first order phase transition."

The first step is to describe the model. Let $A_0 = [0,1]^2$, and for $1 \leq i,j \leq N$, let $B_{ij} = [(i-1)/N, i/N] \times [(j-1)/N, j/N]$. Let $\epsilon_{ij} \in \{0,1\}$ be independent "coin flips" with $P(\epsilon_{ij} = 1) = p$. If $\epsilon_{ij} = 1$, we say B_{ij} is occupied, and we let

$$A_1 = \bigcup_{i,j: \epsilon_{ij}=1} B_{ij},$$

that is, we keep the squares with $\epsilon_{ij} = 1$. To define A_2, we repeat the last construction (appropriately scaled) in each surviving B_{ij}, or more generally, if we have constructed A_{n-1}, then we let $B_{ij}^n = [(i-1)/N^n, i/N^n] \times [(j-1)/N^n, j/N^n]$, $1 \leq i,j \leq N^n$, let $\epsilon_{ij}^n \in \{0,1\}$ be independent with $P(\epsilon_{ij}^n = 1) = p$, and let

$$A_n = A_{n-1} \cap \left(\bigcup_{i,j: \epsilon_{ij}^n=1} B_{ij}^n \right).$$

A_0, A_1, A_2, ... is a decreasing sequence of compact sets, so the limit A_∞ exists (possibly ϕ). Mandelbrot calls A_∞ the curds, and calls the complement $[0,1]^2 - A_\infty$, the whey. Independent of what you call these things, the first question to be resolved is: "When is $A_\infty \neq \phi$?" Using some simple facts about branching processes it is easy to show.

(1) Theorem. $A_\infty \neq \phi$ with positive probability if and only if $p > 1/N^2$.

Proof: Let Z_n be the number of squares of the form B_{ij}^n that are contained in A_n . Z_n is a branching process in which each particle has on the average $N^2 p$ offspring, so if $Np^2 \leq 1$, we have $P(Z_n > 0) \to 0$ as $n \to \infty$, and if $Np^2 > 1$, we have $P(Z_n > 0) \to \rho$ as $n \to \infty$, where ρ is positive solution of

$$((1-p) + p(1-x))^{N^2} = (1-x).$$

(See Athreya and Ney (1972), Chapter 1, for this and other facts we will use below.)

From the results above, we see that if $N^2 p \leq 1$, then $A_n = \phi$ when n is sufficiently large, so $A_\infty = \phi$. On the other hand, if $N^2 p > 1$, then $P(A_n \neq \phi$ for all $n)$ > 0, and since the A_n are a decreasing sequence of nonempty compact sets, we have $A_\infty \neq \phi$ on $\Omega_0 \equiv \{A_n \neq \phi$ for all $n \}$.

Historically, the first aspect of A_∞ that was considered was its "similarity dimension" (See Mandelbrot's Fractal Geometry of Nature (1983), hereafter abbreviated FGN, p. 211.) To calculate this, we observe that:

(i) if we multiply the unit square by N then on the average we have $N^2 p$ copies of our set, and

(ii) if we multiply the unit cube in d–dimensions by N, then we have N^d copies of it.

So the "similarity dimension" of our set is

$$\frac{\log(N^2 p)}{\log N} = 2 + \frac{\log p}{\log N}.$$

(The point of (ii) is that if we apply the last recipe to the unit cube in d dimensions, it gives d.)

The last formula is a well–known recipe for computing the Hausdorff dimension of things (e.g. the standard Cantor set has dimension $\log 2/\log 3$ because multiplying it by 3 produces 2 copies), so it is natural to try to show that the Hausdorff dimension is $2 + (\log p)/\log N$. Half of this is very easy. Well–known results from branching processes (or martingale theory) imply that if Z_n = the number of B_{ij}^n contained in A_n, then

$$W_n = Z_n/(N^2 p)^n \to W \text{ a.s.,}$$

where W is a random variable with $EW = 1$, and

$$\{W > 0\} = \{Z_n > 0 \text{ for all } n\} = \{A_\infty \neq \phi\}.$$

(Again, this can be found in Chapter 1 of Athreya and Ney (1972).)

Since A_n can be covered by $Z_n = W_n \cdot (N^2 p)^n$ cubes with sides of length N^{-n}, we see that if $\alpha = 2 + (\log p)/(\log N)$, then the α–dimensional Hausdorff measure of A_∞ is at most $W < \infty$, so the Hausdorff dimension of $A_\infty \leq \alpha$. To get a bound in the other direction requires more work, but fortunately for us, most of the work has already been done by Kahane and Peyrière (1976) who studied a related random measure. To make connection with their results, consider a process in which the squares B_{ij}^n are assigned weights w_{ij}^n with $P(w_{ij}^n = 1/p) = p$ and $P(w_{ij}^n = 0) = 1-p$; and define a sequence of measures μ_n on $[0,1]^d$ with $\mu_n(dx) = u_n(x)\, dx$, where the u_n satisfy: $u_0 \equiv 1$, and if $x \in B_{ij}^n$

$$u_n(x) = u_{n-1}(x)\, w_{ij}^n.$$

Kahane and Peyrière showed that the support of the limiting measure has Hausdorff dimension α. Since the support of μ_∞ is contained in A_∞, this shows $\dim(A_\infty) = \alpha$.

Note: Mauldin, Graf, and Williams (1987) have investigated Hausdorff dimensions of random recursive constructions which include this one. Their results show that if $h(t) = t^\alpha (\log|\log t|)^{1-(\alpha/2)}$ then the associated Hausdorff measure satisfies $0 < \mathscr{H}^h(A_\infty) < \infty$ when $A_\infty \neq \phi$. This implies in particular that the α–dimensional Hausdroff measure of A_∞ is always 0.

With the random set defined, and its Hausdorff dimension computed, we turn our attention now to our main subject: the connectivity properties of A_∞. The first two results are essentially due to Mandelbrot (see FGN, p. 215–216), but we use branching process arguments instead of his rule that "the co–dimensions add" (FGN, p. 213).

(2) Theorem. If $p \leq 1/N$ and x is not of the form m/N^n for some integers m and n, then $P(A_\infty \cap ([0,1] \times \{x\}) = \phi) = 1$.

Proof: The number of intervals of the form $[(j-1)/N^n, j/N^n] \times \{x\}$ contained in A_n is a branching process in which the mean number of offspring is Np.

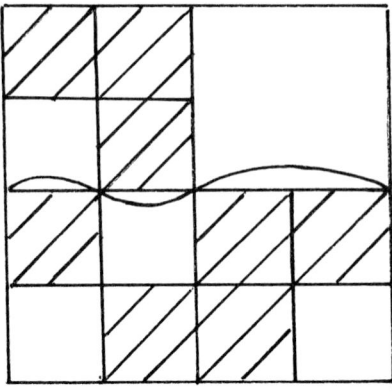

Figure 7.1

The last result (combined with the corresponding fact for vertical lines) implies that if $p \leq 1/N$, then the largest connected component is a point. By changing the value of x that we consider, this result can be improved:

(3) Theorem. If $p \leq 1/\sqrt{N}$ then the largest connected component is a point.

Proof: We say a segment $[(j-1)/N^n, j/N^n] \times \{1/N\}$ is vacant if either of the two adjacent squares in the nth subdivision is. The reason for this terminology can be seen in Figure 7.1, which shows what might happen in the first two subdivisions when $N = 2$. After the first subdivision $[1/2,1] \times \{1/2\}$ was vacant, and after the second the whole interval is. At this point the wiggly line is not a path in the whey, but eventually the squares that touch $(1/4,1/2)$ and $(1/2,1/2)$ will become vacant, and it will be.

With the last picture in mind, we let Y_n be the number of occupied segments of the form $[(j-1)/N^n, j/N^n] \times \{1/N\}$. Y_n is a branching process in which each interval has $p^2 N$ offspring, so if $p^2 N \leq 1$, that is, $p \leq 1/\sqrt{N}$, then the branching process dies out

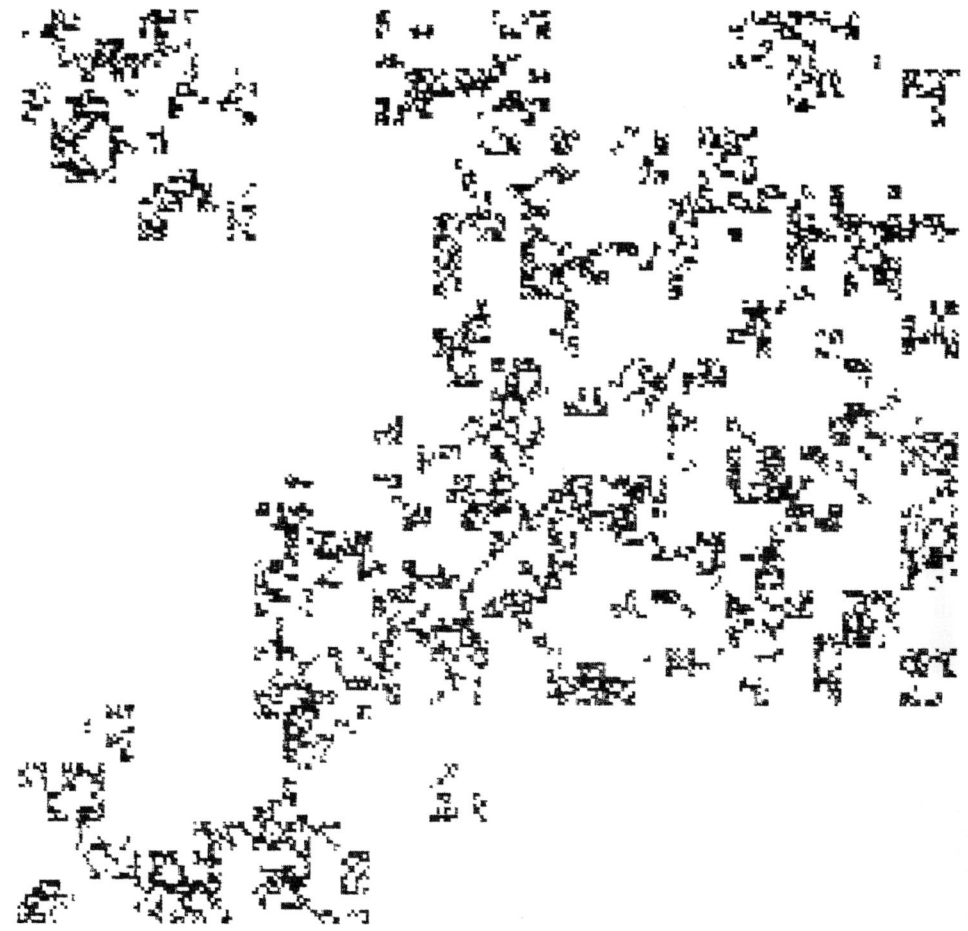

Figure 7.2 P $= 1/\sqrt{2} = .7071$

with probability 1 and, as above, if we wait a while longer there will be a path in the whey arbitrarily close to $[0,1] \times \{1/N\}$. Repeating the last argument at heights k/N^m hows that with probability 1 there are curves in the whey arbitrarily close to all the lines $[0,1] \times \{k/N^m\}$, and combining this with the analogous statement for vertical lines, shows the largest connected component is a point.

Note: When $N = 2$, we have that $A_\infty \neq \phi$ for $p > 1/4$, and (3) applies when $p \leq 1/\sqrt{2}$. A simulation of the first 8 subdivisions for $p = 1/\sqrt{2}$ is given in Figure 7.2.

Having seen that A_∞ can be badly disconnected, the logical next step is to ask if it can ever be connected. If we let $|A_n|$ denote the Lebesgue measure of A_n, then $E|A_n| = p^n \to 0$ exponentially fast, so at first this looks unlikely and in fact, the author once conjectured in public (at a meeting in Frankfurt, Germany) that the conclusion in (3) held for all $p < 1$. This is not the case, however, and a simple argument shows that if p is large enough, then with positive probability there is a connected component that intersects all four sides of the square. After discovering our proof (this is joint work with L. and J.T. Chayes (1988)), we noticed that one of the key ideas appears in Mandelbrot's heuristic argument (see FGN, p. 217), so we will reverse the historical order of things, giving his argument first and using it to motivate our rigorous proof. We invite the reader to put down the book at the end of the next paragraph and fill in the missing details.

"First consider the case in which the number of surviving squares K is nonrandom. In this case, if $N^2 - K > [N/2]$ (where $[x]$ = the largest integer $\leq x$) there is no way that any given face between two precurd cells can fail to survive. Even if the worst happens and all the nonsurviving eddies crowd along said face, these eddies are so insufficient in number that it is sure (not almost, but absolutely) that no path becomes disconnected." Mandelbrot goes on to conclude that: "With the same condition applied to unconstrained curdling, failure to percolate is no longer an

Figure 7.3 P = 8/9

impossibility but an unlikely event." While the last statement is a reasonable conclusion, it is not by mathematical standards a proof (although the bound implied on p_c is probably correct; see Figure 7.3 for a simulation of the system with $p = 8/9$), so we give one.

To state our result, and prepare for other developments below, we need some definitions. Let $B_n = \{x \in A_n : x$ can be connected to $\{0\} \times [0,1]$ and to $\{1\} \times [0,1]$ by paths in $A_n\}$, and let $B_\infty = \cap\, B_n$. If $x \in B_\infty$, let $C_n(x)$ be the component of B_n containing x. $C_1(x) \supset C_2(x) \supset ...$ are compact and connected, so $\Gamma(x) = \cap\, C_n(x)$ is connected and has a nontrivial intersection with $\{0\} \times [0,1]$ and $\{1\} \times [0,1]$ (since all the $C_n(x)$ do). Let $\Omega_1 = \{B_\infty \neq \phi\}$ (and recall $\Omega_0 = \{A_\infty \neq \phi\}$). When Ω_1 occurs, we say there is a left to right crossing of $[0,1]^2$. Let $p_c(N) = \inf\{\, p : P(\Omega_1) > 0\}$.

(4) Theorem. $p_c(N) < 1$ for all $N \geq 2$.

Proof: We will prove the result only in the case $N = 3$. It will be clear from the first observation that the same proof works for any $N \geq 3$. The case $N = 2$ can be treated by comparing with $N = 4$. The first observation explains why we start with $N = 3$. In this case as long as 8 of the 9 squares remain during each subdivision, the set stays connected (even if the worst happens and all the nonsurviving eddies crowd along one face).

This last observation motivates our next set of definitions. We say an outcome is "good" if A_1 contains at least 8 of the squares B_{ij}^1. We say an outcome is "very good" if A_1 contains at least 8 squares B_{ij}^1 that are good; that is, contain at least 8 squares when they are subdivided. For $m \geq 2$, we say an outcome is "$(\text{very})^m$ good" if A_1 contains at least 8 squares B_{ij}^1 in which $A_m \cap B_{ij}^1$ is $(\text{very})^{m-1}$ good.

Let θ_m be the probability that the outcome is $(\text{very})^m$ good. From the recursive definition, it is clear that

$$\theta_m = p^9(\theta_{m-1}^9 + 9\theta_{m-1}^8 (1-\theta_{m-1})) + 9p^8(1-p)\,\theta_{m-1}^8$$

for $m \geq 1$, and

$$\theta_0 = p^9 + 9p^8(1-p).$$

Here, θ_0 is what we get if we let $\theta_{-1} = 1$ in the previous definition, so if we let

$$\varphi(x) = p^9(9x^8 - 8x^9) + 9p^8(1-p)x^8,$$

then $\theta_m = \varphi^{m+1}(1)$, where $\varphi^{m+1}(x) = \varphi(\varphi^m(x)) = \ldots$ A little thought gives

(5) Lemma. As $n \uparrow \infty$, $\varphi^n(1) \downarrow \rho =$ the largest fixed point of φ in $[0,1]$.

With (5) in hand, the proof of (4) will be complete once we show that if p is close to 1, φ has a fixed point > 0. To simplify notation let $\alpha = p^9$, and $\beta = 9p^8(1-p)$, so that φ may be written as

$$\varphi(x) = (9\alpha + \beta)x^8 - 8\alpha x^9$$

Letting $x = 1 - \epsilon$, and observing that

$$(1-\epsilon)^k = 1 - k\epsilon + \frac{k(k-1)}{1 \cdot 2}\epsilon^2 - \frac{k(k-1)(k-2)}{1 \cdot 2 \cdot 3}\epsilon^3 + \ldots,$$

we see that

$$(1-\epsilon)^8 \geq 1-8\epsilon, \quad \text{when } (8-2)\epsilon/3 < 1,$$

and

$$(1-\epsilon)^9 \leq 1-9\epsilon + 36\epsilon^2, \quad \text{when } (9-3)\epsilon/4 < 1.$$

(The indicated conditions imply that the terms we have dropped alternate in sign and decrease in magnitude.)

The last observation implies that for $\epsilon < 2/3$,

$$\varphi(1-\epsilon) \geq (9\alpha+\beta)(1-8\epsilon) - 8\alpha(1-9\epsilon + 36\epsilon^2).$$
$$= (\alpha+\beta) - 8\beta\epsilon - 288\alpha\epsilon^2.$$

Now $\beta \geq 0$ and $\alpha \leq 1$, so if $\epsilon < 1/8$, we have

$$\varphi(1-\epsilon) \geq \alpha - 288\epsilon^2.$$

Setting $\epsilon = .001$ and $\alpha = 1-.5\epsilon$ gives

$$\varphi(1-\epsilon) \geq 1-.788\epsilon,$$

so when $\alpha = .9995$, φ has a fixed point in $[.999,1]$. Recalling $\alpha = p^9$ and $(1-\delta)^9 \geq 1-9\delta$ when $(9-2)\delta/3 < 1$, we see that if $p > .99995$, then $P(\Omega_1) \geq .999 > 0$ proving (4).

At this point, we have seen that if p is ridiculously close to 1, then the probability of a left to right crossing is positive, and that if $p < 1/\sqrt{N}$, then P(largest connected component of A_∞ is a point) $= 1$. It is natural, then, to let

$$p_d = \sup\{ p : P(\text{largest connected component of } A_\infty \text{ is a point }) = 1\},$$
$$p_c = \inf\{ p : P(A_\infty \text{ has a left to right crossing}) > 0\}$$

(where d is for dustlike and c is for critical), and ask if $p_c = p_d$. This, and more, is proved in the next result, which uses the notation introduced before we stated (4).

(6) Theorem. Let $\Omega_1^n = \{B_n \neq \phi\}$. There is an $\epsilon_0 > 0$, so that if $P(\Omega_1^n) \leq \epsilon_0$, then $P(\Omega_1) = 0$, and furthermore, the largest connected component is a point.

The proof of (6) is based on two conclusions that are analogues of results for "ordinary" percolation. The proofs of these results are straightforward generalizations of the "classical" ones, but it will take a lot of verbiage to convince the reader of this, and we will need to prove a second pair of results later in the chapter, so details are deferred to the end.

Let $\Omega_{1,K}^n$ be the event that there is a left–right crossing of $[0,1] \times [0,K]$ when independent copies of A_n are placed in each of the squares $[0,1] \times [k-1,k]$, $1 \leq k \leq K$.

(a) If $P(\Omega_{1,1}^n) \leq \epsilon$, then $P(\Omega_{1,K}^n) \leq f_K(\epsilon)$, where $f_K(\epsilon) \to 0$ as $\epsilon \to 0$.

(b) If $P(\Omega_{1,2}^n) \leq .01$, then $P(\Omega_1^{n+k}{}_{,2}) \leq \frac{1}{25} \exp(-N^{k-1})$.

Remark. You should observe that the last probability goes to 0 exponentially fast in the length of the cubes.

With (a) and (b) in hand, the rest is easy. If we pick ϵ_0 so that $f_2(\epsilon) \leq .01$ for all $\epsilon \leq \epsilon_0$, and have n such that $P(\Omega^n_{1,1}) < \epsilon_0$, then (a) and (b) imply $P(\Omega^n_{1,1}) \leq$ $P(\Omega^n_{1,2})$ goes to 0 as $n \to \infty$. Feeding this estimate into (a), shows $P(\Omega^n_{1,K}) \to 0$ for all $K < \infty$. The last observation implies that with probability 1, we have a "crack" (i.e., a curve from bottom to top which lies completely in the whey) in $[a,b] \times [0,1]$ for all $a < b$ of the form $a = j/N^m$, $b = k/N^m$. Since this will also, with probability 1, be true for all the reflected rectangles $[0,1] \times [a,b]$, it follows that the largest connected component is a point.

From (6), it follows immediately that the percolation probability $P(\Omega_1)$ is positive at p_c. To prove this, note that since $p \to P_p(\Omega^n_1)$ is continuous (it is a polynomial), and $\downarrow P_p(\Omega_1)$ as $n \to \infty$, $p \to P_p(\Omega_1)$ is upper semicontinuous. Since $p \to P_p(\Omega_1)$ is nondecreasing, the last observation implies it must be right continuous on $[0,1]$, and hence positive at p_c; that is, there is positive probability of a left to right crossing when $p = p_c$.

Looking at the last result one might think, "It is easy to see the source of the discontinuity above. Ω_1 is really just a sponge crossing event, so the discontinuity is caused by the phenomenon in (b): if sponge crossing probabilities get too small, then they go to 0." To get around this objection, and have a phase transition which, like other percolation processes, involves the appearance of an unbounded connected component, we place an independent copy of our random set A_∞ in each square $z + [0,1]^2$, $z \in Z^2$, call the result A'_∞, and look for percolation in A'_∞ in the usual sense. We let $\Omega_\infty = \{A'_\infty$ has an unbounded connected component$\}$, and let $p_b = \inf\{p: P(\Omega_\infty) > 0\}$.

It is clear that $p_b \geq p_c \geq p_d$. (This should help explain the somewhat unusual notation. To help you remember what the b stands for, think of unBounded, and observe that p_u stinks.) Our last result shows that $p_b = p_c$.

(7) Theorem. If $p \geq p_c$ then with probability 1, A'_∞ has a unique unbounded connected component.

Comparing this with (6) shows that the system undergoes a very violent transition as we pass through $p = p_c$. When $p < p_c$ the largest connected component is a point, but when $p = p_c$ there is a unique unbounded component. The reader should note that if we let $\Omega_\infty = \{$the unbounded component of A'_∞ touches $[0,1]^2\}$, then $P(\Omega_\infty) > 0$ at p_c, so $p \to P_p(\Omega_\infty)$ is, like $p \to P_p(\Omega_1)$, discontinuous at p_c.

The key to the proof of (7) is the observation that if we rescale A'_∞ by dividing by N, and then flip new coins to see which squares of the form $[(i-1)/N, i/N] \times [(j-1)/N, j/N]$ are occupied, then the result has the same distribution as A'_∞. So if we ignore the second step, we have

$$A'_\infty/N \stackrel{d}{=} (A'_\infty \mid \text{all } \epsilon^1_{ij} = 1),$$

or iterating the last result,

$$A'_\infty/N^n \stackrel{d}{=} (A'_\infty \mid \text{all } \epsilon^m_{ij} = 1, m \leq n) .$$

The last observation makes it easy to believe (and prove) that if $P(\Omega_1) > 0$ then the probability of a left to right crossing of $[0,N^n]$ in A'_∞ approaches 1 as $n \to \infty$. To prove this, we observe that $P(\Omega^n_1) \to P(\Omega_1)$ as $n \to \infty$, so if $\epsilon > 0$ and n is large, then

$$P(\Omega^n_1 - \Omega_1) \leq \epsilon P(\Omega_1) \leq \epsilon P(\Omega^n_1),$$

or

$$P(\Omega_1 | \Omega_1^n) \geq 1-\epsilon,$$

and hence

$$P(\Omega_1 | \epsilon_{ij}^m = 1 \text{ for all } m \leq n) \geq 1-\epsilon.$$

Having established that the crossing probabilities of large squares is close to 1, the result now follows from two more facts we will prove at the end. To avoid the topological nightmares that would come from trying to deal with A_∞' directly, we will consider the situation after n subdivisions, and prove results that are independent of n.

Let A_n' be the set that results when we place independent copies of A_n in each square $z + [0,1]^2$, $z \in Z^2$. Let $\Omega_{J,K}^n$ be the event that there is a left to right crossing of $[0,J] \times [0,K]$ in A_n'.

(a') If $P(\Omega_{L,L}^n) \geq 1-\epsilon$, then $P(\Omega_{kL,L}^n) \geq 1-g_k(\epsilon)$, where $g_k(\epsilon)$ is independent of n and approaches 0 as $\epsilon \to 0$.

(b') If $P(\Omega_{2L,L}^n) \geq .99$, then $P(\Omega_{2^kL,2^{k-1}L}^n) \geq 1 - \frac{1}{25} \exp(-2^{k-1})$.

The notation and the numbers in (b') should remind the reader of (a) and (b) stated earlier. We will see at the end of this chapter that they are closely related. With (a') and (b') in hand, the conclusion follows from the construction we used in the last chapter to produce percolation. Let L be chosen so that $P(\Omega_{2L,L}^n) \geq .99$ for all n when $p = p_c$, and define boxes

$$B_{2j-1} = (0,2^{2j-1}L) \times (0,2^{2j-2}L)$$
$$B_{2j} = (0,2^{2j-1}L) \times (0,2^{2j}L),$$

as we did in Section 6a. The events are chosen so that if we get left to right crossings of all the B_{2j-1}, and top to bottom crossings of all the B_{2j}, then there is an infinite path starting on $\{0\} \times (0,L)$. (See Figure 6.6.) With probability at least

$$\prod_{j=1}^{\infty} [1 - (1/25)\exp(-2^{j-1})] > 0,$$

we get all the crossings we want, so with at least this probability there is an infinite path in A_n' starting at some point in $\{0\} \times [0,L]$. Letting $n \to \infty$, and taking intersections as we did before the statement of Theorem 1, it is easy to see that the same is true for $n = \infty$.

The argument above shows the existence of an unbounded component when $p \geq p_c$. The fact that it is unique is proved in the same way as in the ordinary case. To complete the proofs of (7) and (6), we neeed to prove (a'), (b'), (a), and (b). (Readers who are tired can safely skip this. The arguments will not be needed in what follows.) We will prove them in the order indicated. In each case, our proofs are obtained by modifying one step of the proof of the analogous statements for ordinary percolation, so we will begin by stating those results:

(A)   If $\rho_{L,L} \geq 1-\epsilon$, then $\rho_{kL,L} \geq 1-h_k(\epsilon)$, where $h_k(\epsilon)$ is independent of L and approaches 0 as $\epsilon \to 0$.

(B)   If $\rho_{2L,L} \geq .99$, then $\rho_{2^kL,2^{k-1}L} \geq 1 - \frac{1}{25}\exp(-2^{k-1})$.

(A) is a combination of (2) and (3) in Section 6a, and (B) is (9) from that section. In what follows we will assume you are familiar with the proofs of those results.

(a') If $P(\Omega_{L,L}^n) \geq 1-\epsilon$, then $P(\Omega_{kL,L}^n) \geq 1 - g_k(\epsilon)$, where $g_k(\epsilon)$ is independent of n and approaches 0 as $\epsilon \to 0$.

Proof. From the discussion in Section 6a, it suffices to show that

$$(*) \qquad\qquad\qquad P(F_s | E_s) \geq P(F_s)$$

is valid in our setting. In the independent case, if we condition on the location of s then the sites above s have the same distribution as they did orginally (i.e., independent). In the present setting, this is not true but something better happens. The presence of the path is "good news", i.e., the conditional distribution is larger than the original and the inequality we want is true.

To make the argument in the last paragraph precise we need to introduce some notation to describe the conditional distribution. The reader is encouraged to look at Figure 7.4 while we do this. (In that drawing the shaded squares are occupied, blank squares are vacant (and left unsubdivided on the next level), and squares marked with u are unconditioned; that is, we do not need to know their fate to know that the shaded set is the lowest crossing.)

Let s be the lowest left to right crossing of $[0,1]^2$ in A_n (s = a union of squares) when we consider squares to be adjacent if they share a side in common. Let $\mathscr{A}(s)$ be region above s, defined in the obvious way. A square of the form $[(i-1)/N^m, i/N^m] \times [(j-1)/N^m, j/N^m]$ is said to be unconditioned if it lies in $\mathscr{A}(s)$, because in this case its coin flip ϵ_{ij}^m is independent of the event {s is the lowest left to right crossing}.

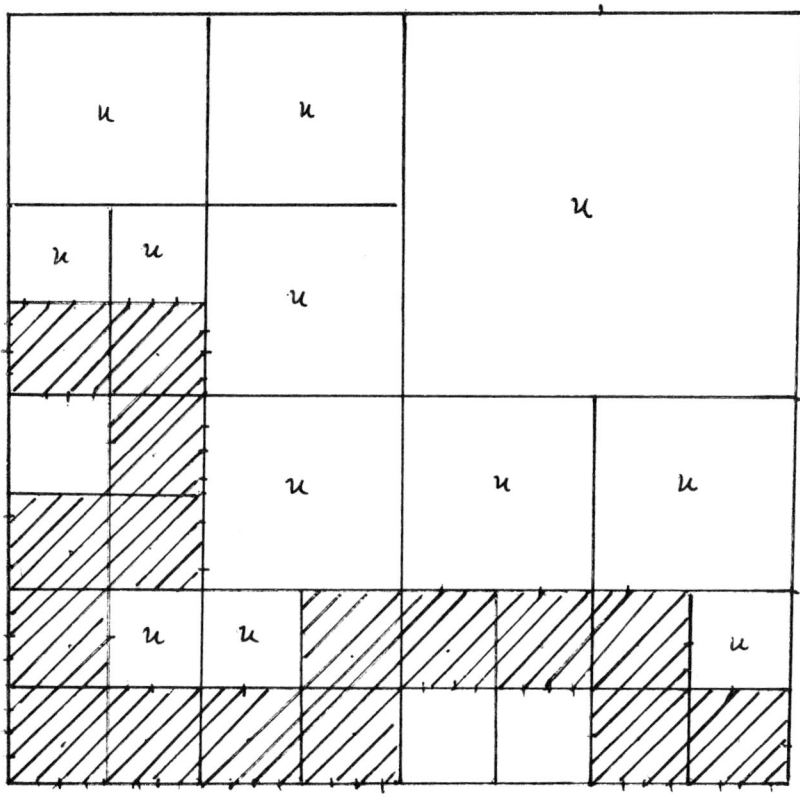

Figure 7.4

In addition to unconditioned squares, there are, of course, also squares that intersect s. The latter squares must be occupied, for otherwise the part of s they touch would not be, so they are our friends. The last observation shows that (∗) holds and completes the proof of (a'), so we proceed now to the proof of:

(b') If $P(\Omega^n_{2L,L}) \geq .99$, then $P(\Omega^n_{2^kL,2^{k-1}L}) \geq 1 - \frac{1}{25}\exp(-2^{k-1})$.

Proof: From our discussion of the independent case, it suffices to show that (7) in

Section 6a is valid in this setting, but this is trivial. The existence of a crossing in

[0,4L] × [0,L] and [0,4L] × [L,2L] ARE independent events.

Having dealt with (a') and (b'), the next item on the agenda is to prove:

(a) If $P(\Omega^n_{1,1}) \leq \epsilon$, then $P(\Omega^n_{1,K}) \leq f_k(\epsilon)$, where $f_k(\epsilon) \to 0$ as $\epsilon \to 0$.

Proof: The first step is to turn this into a problem about percolation probabilities close to 1, by looking at the vacant sites on the dual graph $G^* = (\mathbb{Z}^2, \mathscr{E}^*)$, that is, the points in the graph are $\mathbb{Z}^2$ and the edges $\mathscr{E}^* = \{(z,z+u) : u_1^2 + u_2^2 \leq 2\}$ i.e., in addition to nearest neighbor connections $z \to z+(1,0)$, ... $z \to z+(0,-1)$, there are connections to diagonal nearest neighbors $z \to z+(1,1)$, ... $z \to z+(1,-1)$.

   G^* is the (graph–theoretic) dual of $G = (\mathbb{Z}^2, \mathscr{E})$ where $\mathscr{E} = \{(z,z+u) : u_1^2 + u_2^2 \leq 1\}$ is the usual set of edges connecting nearest neighbor sites, and, as is well known, either there is always either an occupied left to right crossing of $\{1,...J\} \times \{1,...K\}$, or a vacant top to bottom crossing of the same rectangle, but not both. (See Section 6b for more details.) From the last observation, it follows that if we let $\Omega^n_{J,K}$ be the probability of a top to bottom crossing of $[0,J] \times [0,K]$ by vacant squares when we count all squares that touch as adjacent, then it is sufficient to prove

(ã) If $P(\Omega^n_{1,1}) \geq 1-\epsilon$, then $P(\Omega^n_{1,K}) \geq 1-f_K(\epsilon)$, where $f_K(\epsilon) \to 0$ as $\epsilon \to 0$.

Proof: The proof of the last result is very similar to the proof of (a'). From our discussion of the independent case it suffices to show that

(*) $P(\check{F}_s | \check{E}_s) \geq P(\check{F}_s)$

where the $\sim$ indicates that we are referring to vacant crossings, but otherwise considering the same events as in the proofs of (2) in Section 6a, and (a') above.

To prove ($*$) this time, we repeat the argument in the proof of (a'). The unconditioned squares are still unconditioned, and the ones that touch s (now a vacant crossing) are affected by knowing the square they touch is vacant, but this time we cannot conclude that the corresponding coin flip $= 0$. It is a delicate matter to prove (and in general false) that conditioning on an increasing event causes the set of vacant sites to be larger than the original in the sense of stochastic monotonicity (i.e., the two sets can be constructed on the same space in such a way that one includes the other). Fortunately we do not need this here. We are interested in only one decreasing event, so it follows from Harris' inequality that

$$P(\tilde{F}_s | \tilde{E}_s) \geq P(\tilde{F}_s),$$

and the proof of (a) is complete.

The last thing to be shown is

(b) If $P(\Omega_{1,2}^n) \leq .01$, then $P(\Omega_{1,2}^{n+k}) \leq \frac{1}{25} \exp(-N^{k-1})$.

Proof: As in the proof of (a), it suffices to prove

($\tilde{b}$) If $P(\tilde{\Omega}_{1,2}^n) \geq .99$, then $P(\tilde{\Omega}_{1,2}^n) \geq 1 - \frac{1}{25} \exp(-N^{k-1})$.

Now

$$P(\tilde{\Omega}_{N^m,2N^m}^n) = P(\tilde{\Omega}_{1,2}^{n+m} \mid \text{all } \epsilon_{ij}^k = 1 \text{ when } k \leq m),$$

so it suffices to show

$$P(\Omega^n_{N^m, 2N^m}) \to 1,$$

exponentially fast. Having changed our perspective so that the squares get larger, the adjacent rectangles used in the proof of (7) in Section 6a ARE independent, and repeating the proof of (b') proves ($\check{b}$) and hence (b).

8 First Passage Percolation

8a Limit Theorems for First Passage Times

In (1965), Hammersley and Welsh introduced the following model of the spread of a fluid through a porous medium. Consider $\mathbb{Z}^2$ as a graph, with edges connecting each pair of points with $|x-y| = 1$. With edge e (which we will also call a bond), there is associated an independent nonnegative random variable X(e) with distribution F, which represents the amount of time it takes the fluid to flow through the opening that the edge represents. If $x,y \in \mathbb{Z}^2$, let t(x,y) be the minimum travel time over all paths from x to y. Here a path is, as usual, a sequence of points $x_0 = x, x_1, \dots x_n = y$ with $|x_{m-1}-x_m| = 1$ for $1 \le m \le n$, and the travel time for such a path is $X(x_0,x_1) + \dots X(x_{n-1},x_n)$. t(x,y) is the time fluid will first appear at y if there is a source at x that begins operating at time 0. The main aim of first passage percolation is to determine the limiting behavior of t(0,x) as $x \to \infty$.

If the times on the bonds have an exponential distribution $F(x) = (1- e^{-\lambda x})$ for $x \ge 0$, then the model above is equivalent to the growth model ξ_t discussed in Chapter 1. To make the connection, let $s(x) = \inf \{ t : x \in \xi_t^0 \}$ (where ξ_t^0 indicates the state at time t starting from a single occupied site at 0), and let X(x,y) be the amount of time we have to wait for the next arrow from x to y; that is, in the notation of Chapter 1,

$$X(x,y) = \inf \{ T_n^{x,y} - s(x) : T_n^{x,y} > s(x) \}.$$

A little thought reveals that the X(x,y) are i.i.d., and that s(x) is the first passage time t(0,x) defined above.

In Chapter 1, we used the subadditive ergodic theorem to prove a "shape theorem" for the growth model. To do this, we first extended $t(0,x)$ to $x \in \mathbb{R}^2$ by letting $t(0,x)$ be the passage time to the closest point in $\mathbb{Z}^2$ (and taking the minimum $t(0,z)$ if several z are tied for this honor). Second, we showed that for each $x \in \mathbb{R}^2$ there is a constant $\mu(x)$ so that as $n \to \infty$

$$t(0,nx)/n \to \mu(x) \quad \text{a.s.}$$

Finally, we proved that if

$$\bar{\xi}_t^0 = \{ \, x \in \mathbb{R}^2 : t(0,x) \le t \, \},$$

then there is a convex set $A = \{ \, x : \mu(x) \le 1 \, \}$ so that for any $\epsilon > 0$

$$P(\, (1-\epsilon)tA \subset \bar{\xi}_t^0 \subset (1+\epsilon)tA \,) \to 1 \quad \text{as } t \to \infty.$$

In this section, we prove the analogous statements for first passage percolation. The first result gives the limiting behavior of the first passage times $t(0,nx)$. For simplicity, we will state and prove the result only for $x = e_1 = (1,0)$. If we extend $t(0,y)$ to $y \in \mathbb{R}^2$ using the rule described above, then the same result holds with e_1 replaced by any $x \in \mathbb{R}^2$, and μ by $\mu(x)$.

(1) Theorem. Let $Y = \min (X_1,X_2,X_3,X_4)$ where the X_i are independent and have distribution F.

(a) There is a constant $\mu < \infty$ so that

$$t(0,ne_1)/n \to \mu \quad \text{a.s. as } n \to \infty,$$

if and only if $E(Y) < \infty$.

(b) For any distribution F, there is a constant $\mu < \infty$ so that

$$t(0,ne_1)/ \, n \to \mu \text{ in probability as } n \to \infty,$$

and

$$\liminf_{n \to \infty} t(0,ne_1)/n = \mu \quad \text{a.s.}$$

Proof of (a): To prove the necessity, let

$$T_n = \min \{ \, t(e) : e \text{ is an edge with one end at } ne_1 \, \},$$

observe that T_n has the same distribution as Y, and that $T_2, T_4, T_6, \ldots$ are independent. So if $E(Y) = \infty$,

$$\sum_{n=0}^{\infty} P(T_{2n} \geq 2Kn \,) \geq (1/2K) \int_0^{\infty} P(Y \geq y \,) \, dy = \infty,$$

and it follows from the Borel–Cantelli lemma that

$$K \leq \limsup_{n \to \infty} T_{2n}/2n \leq \limsup_{n \to \infty} t(0,ne_1)/n.$$

Since the last conclusion holds for any $K < \infty$, $t(0,ne_1)/n$ cannot converge to a finite limit when $E(Y) = \infty$. (To complete the picture you should note that it follows from (b) that $t(0,ne_1)$ does not converge to ∞ either.)

To prove that $E(Y) < \infty$ is sufficient for convergence, we define four disjoint paths from (0,0) to (1,0):

$r_1 : (0,0) \rightarrow (1,0)$

$r_2 : (0,0) \rightarrow (0,1) \rightarrow (1,1) \rightarrow (1,0)$

$r_3 : (0,0) \rightarrow (0,-1) \rightarrow (1,-1) \rightarrow (1,0)$

$r_4 : (0,0) \rightarrow (-1,0) \rightarrow (-1,1) \rightarrow (-1,2) \rightarrow (0,2) \rightarrow (1,2) \rightarrow (2,2) \rightarrow (2,1) \rightarrow (2,0) \rightarrow (1,0)$

If we let $t(r_i)$ be the travel time for path r_i, then

$$P(\, t(0,e_1) \geq t \,) \leq P(\, \min_{1 \leq i \leq 4} \, t(r_i) \geq t \,) = \prod_{i=1}^{4} P(\, t(r_i) \geq t \,)$$

$$\leq P(\, t(r_4) \geq t \,)^4 \leq (\, 9\, P(\, X \geq t/9 \,))^4,$$

since $t(r_4) \geq t$ only if one of the nine bonds has travel time $\geq t/9$. Since $P(\, Y \geq t \,) = P(\, X \geq t \,)^4$, integrating the last result from 0 to ∞ shows $E\, t(0,e_1) < \infty$. This means that the subadditive ergodic theorem in Chapter 1 can be applied to conclude that as $n \rightarrow \infty$

$$t(0,ne_1)/n \rightarrow \inf_{m} \, Et(0,me_1)/m \quad \text{a.s.}$$

Proof of (b): When $E(Y) = \infty$, all the passage times $t(0,x)$ have infinite expected value, since they are all larger than the time it takes to escape from $0 = \min \{\, t(e) : e$ has one endpoint at $0 \,\}$. To prove (b), we replace $t(x,y)$ by another, almost subadditive, process $\hat{t}(x,y)$ which has $E\,\hat{t}(x,y) < \infty$. Intuitively, $\hat{t}(x,y)$ is the time to go from a point near x to another point near y. To define $\hat{t}(x,y)$, we pick M so that $F(M) = p >$

1/2, call bonds with travel time $\leq$ M open, and those with travel time $>$ M closed. From results in Chapter 6 it follows that there is a unique infinite connected set of open bonds, and for each x there is an open circuit that is part of the infinite cluster and contains x in its interior. Let $\Delta(x)$ be the minimal open circuit with these properties.

The minimal circuit is self–avoiding (i.e., it visits each point at most once), and viewed as a curve in $\mathbb{R}^2$ divides the plane into two connected components. Let $\Delta^0(x)$ denote the component that is bounded (colloquially called the inside), and let $\overline{\Delta}(x) = \Delta(x) \cup \Delta^0(x)$. (The bar is to suggest "closure.") Finally let $\overset{\vee}{\Delta}(x)$ consist of the bonds on $\Delta(x)$, and all the bonds in $\Delta^0(x)$ that are part of the infinite cluster of open bonds. The reason for the last definition will become clear as the proof proceeds. Let

$$\hat{t}(x,y) = \inf\{\ t(x',y') : x' \in \overline{\Delta}(x),\ y' \in \overline{\Delta}(y)\ \},$$

$$u(x) = \sum_{e \in \overline{\Delta}(x)} X(e), \quad \text{and} \quad v(x) = \sum_{e \in \overset{\vee}{\Delta}(x)} X(e).$$

The next result relates $\hat{t}(x,y)$ to $t(x,y)$.

(2) $$\hat{t}(x,y) \leq t(x,y) \leq u(x) + \hat{t}(x,y) + u(y).$$

Proof: The first inequality is trivial since $x \in \overline{\Delta}(x)$ and $y \in \overline{\Delta}(y)$. To see the second, observe that if r is a path from $\overline{\Delta}(x)$ to $\overline{\Delta}(y)$, then $t(x,y) \leq u(x) + t(r) + u(y)$, where $t(r)$ is the travel time along the path r. Taking the infimum over r gives the desired result.

From similar reasoning we see that $\hat{t}(x,y)$ is almost subadditive:

(3) $\hat{t}(x,z) \leq \hat{t}(x,y) + v(y) + \hat{t}(y,z).$

Proof: If r is a path from $\overline{\Delta}(x)$ to $\overline{\Delta}(y)$ and s is a path from $\overline{\Delta}(y)$ to $\overline{\Delta}(z)$, then r and s can be connected by $\check{\Delta}(y)$. Since the sum of all the travel times on $\check{\Delta}(y)$ is $v(y)$, taking the infimum over r and s gives the desired result.

Wait a minute! Were you fooled by the proof just given? A little puzzled by why a ˜ is needed on $\Delta(y)$? In reading the proof, you probably drew a mental picture like the top drawing in Figure 8.1. Such "proofs by picture" are dangerous, since the bottom drawing is also a possiblity and will be a recurring nightmare below. The result is true

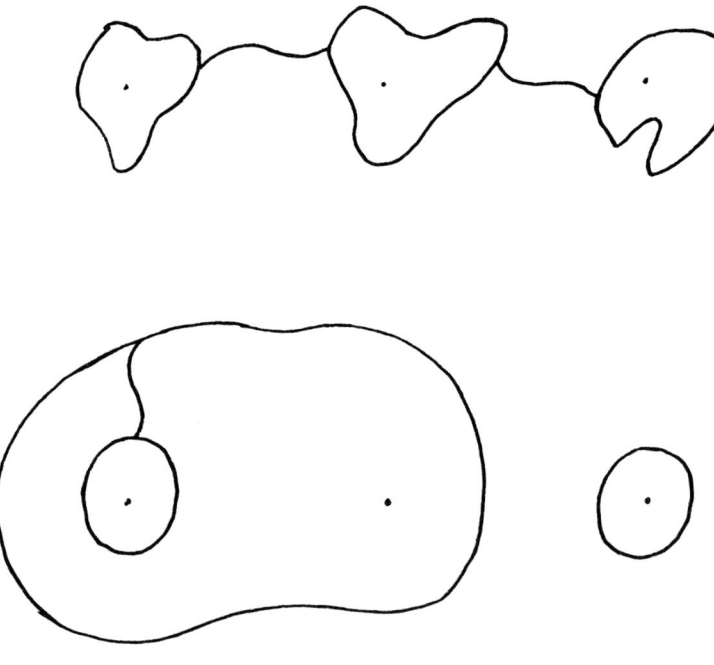

Figure 8.1

in this case, of course. $\hat{t}(x,y) = 0$ but thanks to including all the bonds in $\Delta^0(x)$ which
are part of the infinite cluster, $v(y) + \hat{t}(y,z) \geq \hat{t}(x,z)$.

We leave it to the reader to construct a complete proof of (3) by observing that
either $\overline{\Delta}(x) \cap \overline{\Delta}(y) = \phi$ or $\neq \phi$, and combining this with the two possibities for $\overline{\Delta}(y) \cap$
$\overline{\Delta}(z)$ gives four cases, two of which have been treated above. In a minute, we will show

$$(4) \qquad\qquad E \; \hat{t}(x,y) < \infty \;\; \text{and} \;\; E \; v(x) < \infty.$$

To motivate the reader for this technicality, we will now show that (2), (3), and (4)
imply the desired result. The first step is to add $v(z)$ to both sides of (3) to get

$$\hat{t}(x,z) + v(z) \leq \hat{t}(x,y) + v(y) + \hat{t}(y,z) + v(z).$$

The last inequality and (4) imply that

$$s_{m,n} \equiv \hat{t}(me_1,ne_1) + v(ne_1)$$

is a process to which the subadditive ergodic theorem can be applied, and we conclude
that as $n \to \infty$

$$s_{0,n}/n \to \inf_{m} E \; s_{0,m}/m \;\; \text{a.s.}$$

To get from the last result to a conclusion about $\hat{t}$, we observe that since $v(ne_1)$ are
i.i.d. with finite mean, the Borel–Cantelli lemma implies that for any $\epsilon > 0$

$$\sum_{n=1}^{\infty} P(\ v(ne_1) \geq n\epsilon\ ) \leq (1/\epsilon) \int_0^{\infty} P(\ v(0) \geq x\ )\ dx < \infty,$$

and since ϵ is arbitrary, it follows that as $n \to \infty$

$$\hat{t}(0,ne_1)/n \to \inf_m E\ s_{0,m}/m \ \text{ a.s.}$$

To translate this into a result about t, we observe that (2) implies

$$\hat{t}(0,ne_1) \leq t(0,ne_1) \leq u(0) + \hat{t}(0,ne_1) + u(ne_1).$$

Dividing both sides by n and observing that the distribution of $u(ne_1)$ is independent of n, we see that as $n \to \infty$, $u(0)/n \to 0$ on $\{u(0) < \infty\}$, $u(ne_1)/n \to 0$ in probability, and

$$\liminf_{n \to \infty} u(ne_1)/n = 0 \ \text{a.s},$$

so the proof of (1) will be complete when we show (4).

To prove (4), we begin by observing that results in Chapter 6 imply that if $p > 1/2$, the probability of a left to right crossing of a 6L × 2L sponge, $\rho_{6L,2L}$, approaches 1 as $L \to \infty$. Now left to right crossings of $(-3L,3L) \times (L,3L)$ and $(-3L,3L) \times (-3L,-L)$, and top to bottom crossings of $(-3L,-L) \times (-3L,3L)$ and $(L,3L) \times (-3L,3L)$ can be combined to produce a circuit around 0. Harris' inequality implies that all four circuits exist with a probability at least $(\rho_{6L,2L})^4$, so if we let $L = 3^n$, $n = 1,2,...$, and let ρ_n denote the value of $(\rho_{6L,2L})^4$ when $L = 3^n$, then

(5) $$P \left(\Delta(0) \text{ touches } (-3^{n+1}, 3^{n+1})^c \right) \leq \prod_{m=1}^{n} (1-\rho_m).$$

The last result gives us most of what we need to prove (4). The other ingredient is the following simple calculation:

(6) Lemma. Let Z be a nonnegative random variable such that $P(Z \geq b^k) \leq Ac^k$ for all $k \geq 0$. If $b^m c < 1$ then $E(Z^m) < \infty$.

Proof: Breaking the sum into blocks $b^k \leq n < b^{k+1}$, $k = 0,1,2,...$ shows

$$\sum_{n=1}^{\infty} n^{m-1} P(Z \geq n) \leq \sum_{k=0}^{\infty} b^{(k+1)} \cdot b^{(k+1)(m-1)} \cdot Ac^k = Ab^m \sum_{k=0}^{\infty} (b^m c)^k < \infty.$$

If we let $|\overline{\Delta}(x)| = $ the number of bonds in $\overline{\Delta}(x)$, then (5) and (6) imply that $E |\overline{\Delta}(x)|^m < \infty$ for all $m < \infty$, and since $|v(x)| \leq M|\overline{\Delta}(x)|$, it follows that $E(v(x)^m) < \infty$ for all $m < \infty$. To prove that the same conclusion holds for $\hat{t}(x,y)$, let $z_0 = x$, z_1, ... $z_n = y$ be a path from x to y, that is, $|z_m - z_{m-1}| = 1$ for $1 \leq m \leq n$, and observe that

(7) Lemma. There is a path from $\overline{\Delta}(x)$ to $\overline{\Delta}(y)$ in the union of the $\breve{\Delta}(z_m)$ $1 \leq m \leq n$.

Proof: If you believe this is true, you can skip the proof. I have supplied the details only to save you the trouble of constructing your own. It suffices to show that if $|u-v| = 1$ then $\breve{\Delta}(u) \cap \breve{\Delta}(v) \neq \phi$. If $\Delta^0(u) \subset \Delta^0(v)$, then by definition $\breve{\Delta}(u) \subset \breve{\Delta}(v)$. (This is why we include all the bonds in the interior that are part of the infinite cluster.) Now $\Delta^0(u) \cap \Delta^0(v) \neq \phi$ (since $|u-v| = 1$ and u and v are in the interior of their respective Δ's), so if $\Delta^0(u)$ is not $\subset \Delta^0(v)$ then there are points a,b in $\Delta^0(u)$ so that $a \in \Delta^0(v)$

and b $\notin \Delta^0(v)$. Since a,b $\in \Delta^0(u)$ they can be connected by a polygonal curve which lies entirely in $\Delta^0(u)$. Since a $\in \Delta^0(v)$ and b $\notin \Delta^0(v)$, the curve must intersect $\Delta(v)$ at some point c. Since c $\in \Delta(v)$, c $\in$ e for some edge e $\subset \breve{\Delta}(v)$. c is in $\Delta^0(u)$ and e $\subset \breve{\Delta}(v) \subset$ the infinite cluster, so e is in $\breve{\Delta}(u)$, and we have shown that $\breve{\Delta}(u)$ and $\breve{\Delta}(v)$ intersect.

With the last observation in hand, the conclusion $E(\hat{t}(x,y)^m) < \infty$ follows from the fact that

$$\hat{t}(x,y) \leq M \sum_{m=0}^{n} |\breve{\Delta}(z_m)| \leq M \sum_{m=0}^{n} |\overline{\Delta}(z_m)|$$

This completes the proof of (4), and hence of (1).

With (1) proved, our next goal is to prove a "shape theorem" for

$$\overline{\xi}_t = \{ x \in \mathbb{R}^2 : t(0,x) \leq t \}.$$

The next result gives necessary and sufficient conditions for the desired conclusion to hold. Notice that the answer is in terms of the same random variable that appears in (1), and again a weaker version holds for any distribution F. To keep things simple we have imposed an unnecessary condition on F: $F(0) = 0$. This guarantees that the limit μ in (1) is > 0. At the end of this section and in the next one, we will discuss what happens when there is an atom at 0.

(8) **Theorem.** Suppose $F(0) = 0$. Let $Y = \min(X_1, X_2, X_3, X_4)$ where the X_i are independent and have distribution F.

(a) There is a convex set A so that for any $\epsilon > 0$

$$P \left((1-\epsilon)tA \subset \overline{\xi}_t \subset (1+\epsilon)tA \right) \to 1 \text{ as } t \to \infty$$

if and only if $E(Y^2) < \infty.$

(b) For any distribution F, there is a convex set A so that as $t \to \infty$

$$P \left(\overline{\xi}_t \subset (1+\epsilon)tA \right) \to 1,$$

and

$$|(t^{-1}\overline{\xi}_t) \vartriangle A| \to 0 \text{ in probability.}$$

(Here $\vartriangle$ denotes symmetric difference, and $|S|$ denotes the Lebesgue measure of S.)

The second conclusion says that, although $\overline{\xi}_t$ may not fill up all of $(1-\epsilon)tA$, it fills up most of tA. To show why we are interested in a result like (b), we will derive two corollaries before we prove it. To do this, we need to know that the limit set $A = \{ x : \mu(x) \leq 1 \}$, where $\mu(x)$ is the limit (in probability) of $t(0,nx)/n$. Our first corollary concerns the "x reach process" introduced by Smythe and Wierman (1978). (See their Theorem 6.3.) Projecting $t^{-1}\overline{\xi}_t$ and $A = \{ x : \mu(x) \leq 1 \}$ onto the x axis, and observing that A is convex and invariant under the transformation $(x,y) \to (x,-y)$ (reflection through the x axis), we get

(9) Corollary. Let $x_t = \sup \{m : t(0, (m,n)) \leq t \text{ for some } n \in \mathbb{Z} \}$. As $t \to \infty$

$$x_t/t \to 1/\mu(e_1) \quad \text{a.s.}$$

If we regard the x reach process sample paths as functions, then the inverse functions $b_m = \inf \{ t(0, (m,n)\,) : n \in \mathbb{Z} \}$ are the "point to line process" introduced by Hammersley and Welsh (1965). From (9), we get one of the basic results about this process.

(10) Corollary. As $m \to \infty$, $b_m/m \to \mu(e_1)$ a.s.

Note that if $a_m = t(0,me_1)$, then $a_m/m \to \mu(e_1)$ in probability (and almost surely if $E(Y) < \infty$), so that except for the difference in the type of convergence, the point to point and point to line passage times grow at the same rate. You should also note that (10) holds for any distribution. (This is true without the assumption $F(0) = 0$. See the remark at the end of the section.). Wierman and Reh (1978) proved the result for F with finite mean. At first glance, it might seem that their result should be a special case of the subadditive ergodic theorem. It is not, because no one knows how to extend the sequence b_m to a stationary subadditive process. (Can you find a way to do this? I doubt it, but you should stop for a moment and think about the problem.)

Proof of (a): As before, the necessity of $E(Y^2) < \infty$ is easy. Let $Y(z)$ be the minimum of the travel times for the four bonds that contain z. For any $K < \infty$, we have

$$\sum_{x \in (2\mathbb{Z})^2} P(\, Y(z) \geq K|z| \,) \geq \sum_{n=0}^{\infty} 4n\, P(\, Y(0) \geq Kn \,) \geq 2E(Y/K)^2 = \infty.$$

Since $t(0,z) \geq Y(z)$, it follows from the Borel–Cantelli lemma that

$$P(\, \sup_{z \neq 0} t(0,z)/|z| = \infty \,) = 1,$$

so (a) is false if $E(Y^2) = \infty$.

The proof that $E(Y^2) < \infty$ is sufficient for (a) is very similar to the proof of (b). Because of this, we will only sketch the proof and refer the reader to Section 3 of Cox and Durrett (1981) for the missing details. The sketch below gives the outline of the proof of (b), so even if for some reason you are only interested in the proof of (b), you should still read the next four paragraphs.

We begin by constructing four disjoint paths from $(0,0)$ to $(5,0)$ which are similar to the four paths used in the proof of part (a) of (1). Here $x \to y$ means connect x and y by a straight line path:

$r_1 : (0,0) \to (5,0)$

$r_2 : (0,0) \to (0,1) \to (5,1) \to (5,0)$

$r_3 : (0,0) \to (0,-1) \to (5,-1) \to (5,0)$

$r_4 : (0,0) \to (-1,0) \to (-1,-2) \to (6,-2) \to (6,0) \to (5,0)$

By repeating the argument given in the proof of part (a) of (1), it is easy to show that $E(t(0,5e_1)^2) < \infty$. We could have proved the last conclusion, with e_1 in place of $5e_1$, by using the four old paths from $(0,0)$ to $(1,0)$. We chose to go from $(0,0)$ to $(5,0)$, so that the four paths would fit in the union of the disjoint retangles $(0,0) + (-5/2,5/2]^2$ and $(5,0) + (-5/2,5/2]^2$. The last fact simplifies the computations below.

By translation and rotation through a multiple of 90 degrees, we can extend the construction in the last paragraph to any points $5x$ and $5y$ where $x,y \in \mathbb{Z}^2$ and $|x-y| = 1$. Let $\tau(x,y)$ be the minimum passage time between $5x$ and $5y$ over the four paths. If $z \in \mathbb{Z}^2$, and $z_0 = 0, z_1, \ldots z_n = z$ is a path from 0 to z with $n = |z|$, then subadditivity implies

$$t(0,5z) \leq \sum_{m=1}^{n} t(5z_{m-1}, 5z_m) \leq \sum_{m=1}^{n} \tau(5z_{m-1}, 5z_m).$$

If we let T_m be the mth term in the sum, then T_j and T_k are independent when $|j-k|$ > 1 (here we use the observation about disjoint rectangles), so the variance of the last sum $\leq$ Cn. If we let $K = 2 \, ET_m$, and use Chebyshev's inequality, it follows that

(i) $P(\, t(0,5z) > K|z|\,) \leq C/|z|.$

Here, and in what follows, C and K will denote constants $< \infty$ which will change from line to line. In some cases we will use C' to emphasize that the constant has changed.

The last bound is not very good, but can be improved by considering three disjoint paths from 0 to z. If we assume that $z_1, z_2 \geq 0$, then the paths are:

$$r_1 : (0,0) \to (z_1,0) \to (z_1,z_2)$$
$$r_2 : (0,0) \to (0,1) \to (z_1-1,1) \to (z_1-1,z_2) \to (z_1,z_2)$$
$$r_3 : (0,0) \to (0,-1) \to (z_1+1,1) \to (z_1+1,z_2) \to (z_1,z_2)$$

where again $x \to y$ indicates moving from x to y by the straight line path. (The paths to points in other quadrants can be obtained by a clockwise rotation of the case treated above.) Since the τ's which correspond to steps in one path are, except for the first and last steps, independent of the τ's in the other two paths, a little work results in a bound of the form:

(ii) $P(\, t(0,x) \geq K|x|\,) \leq C/|x|^3 + \epsilon(|x|)$

where $C, K \in (0, \infty)$ and $\Sigma |x| \ \epsilon(|x|) < \infty$. (The second term comes from estimating the probability the sum of the τ's for the first and last steps are $> |x|$.) For more details see (13)–(16) below, or Cox and Durrett (1981), Lemma 3.3.

The last inequality gives us a bound for $P(\ t(0,x) \geq K|x| \ )$ which can be summed over all $x \in \mathbb{Z}^2$ to conclude that

(iii) $P \ (\ \overline{\xi}_t \supset D(0, t/K)\ ) \to 1$ as $t \to \infty$

where $D(0, r) = \{\ x : |x| \leq r\ \}$. With this and a limit theorem for the passage times $t(0, nx)$ in hand, the Denouement of Chapter 1 takes over, and completes the proof of (a).

To prove (b), we begin by proving a shape theorem for $\hat{\xi}_t = \{\ x \in \mathbb{R}^2 : \hat{t}(0,x) \leq t\ \}$. (Here, to economize on notation, we are assuming that $\hat{t}(0,x)$ has already been extended to $x \in \mathbb{R}^2$ in the usual way, so $\hat{\xi}_t$ is a solid blob.) The keys to the proof of (a) above were (i) and (ii). Our first goal is to prove similar results for $\hat{t}$. To do this, we will again construct several disjoint paths from 0 to x. The paths we construct for this proof will be slightly different, since we have no control over how large the circuits $\hat{\Delta}(x)$ that appear in the definition of $\hat{t}(x,y)$ are. A second, more serious, problem is that if γ is a self–avoiding circuit, then the event $\{\ \hat{\Delta}(x) = \gamma\ \}$ is not measurable with respect to the bonds inside γ, since γ must be part of the infinite cluster. To avoid this problem, we will abandon the $\hat{\Delta}(x)$ in favor of the circuits that we will now define.

Let $c(x)$ be the minimal self–avoiding circuit of open bonds that contains x in its interior, and let $\tilde{c}(x)$ be the set of bonds that are on $c(x)$ or in its interior, $c^0(x)$, and part of the infinite cluster. To get estimates on the travel time from 0 to z, we will construct special paths from 0 to z using the following:

(11) Lemma. Let $z_0, \ldots z_n$ be a self–avoiding path from 0 to z. Then the union of the $\tilde{c}(z_m)$ $0 \leq m \leq n$ contains a path from $\overline{\Delta}(0)$ to $\overline{\Delta}(z)$.

Proof: As in the case of (7), if you believe this is true, you can skip the proof. It is included on the theory that it is less painful to read than to discover for yourself. Let $I = \{\, i : c^0(z_i)$ is not contained in any other $c^0(z_j)$ $j \neq i \,\}$. It is easy to see that all the z_m $0 \leq m \leq n$ are contained in some $c^0(z_i)$ with $i \in I$. Let i_0 = the smallest $i \in I$ so that $0 \in c^0(z_i)$. If you have picked $i_0,\ldots,i_k$ then to pick the next index, find the largest j so that $z_j \in G_k \equiv$ the union of the $c^0(z_i)$ $i \in \{\, i_0, \ldots i_k \,\}$. If $j = n$ we are done. If not, let i_{k+1} be the smallest i so that $i \in I$ and $z_{j+1} \in \overline{c}(z_i)$ $(\, = c^0(z_i) \cup c(z_i) \,)$.

Figure 8.2 indicates one of the possible outcomes of the construction. (Readers with a keen eye for detail may object that the circuits drawn are not minimal, but we do not take advantage of the minimality in the proof.) No matter how weird things

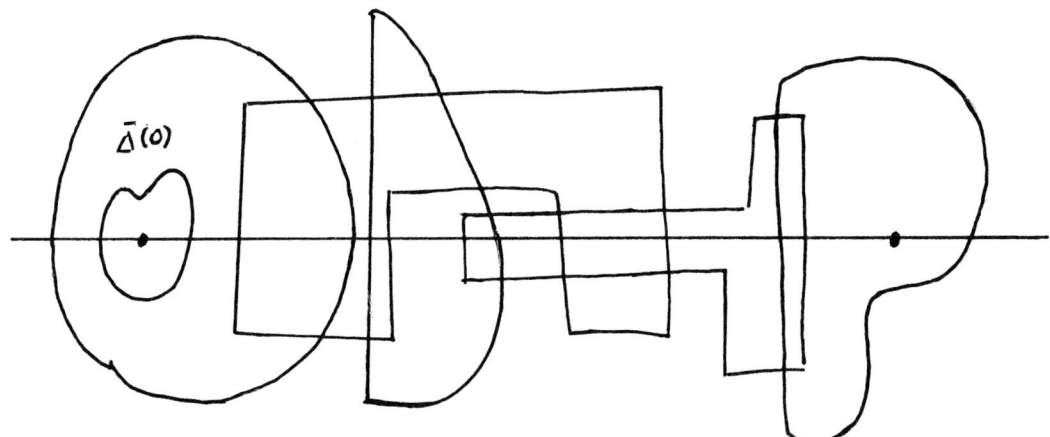

Figure 8.2

get, the new circuit must always intersect the last term in the sequence. To prove this let $w = z_{i(k)}$. By the way points are chosen, the new point z_{j+1} is on the boundary of $\bar{c}(w)$. If i is the newly chosen index, the last observation implies $c^o(w) \cap c^o(z_i) \neq \phi$. The definition of I implies that neither $c^o(z_i)$ nor $c^o(w)$ contains the other, so repeating the proof of (7) shows that $\tilde{c}(w)$ and $\tilde{c}(z_i)$ must intersect.

The last fact implies that the union of the circuits chosen is connected. The last detail is to show that it intersects $\overline{\Delta}(0)$ (and $\overline{\Delta}(z)$). If $c(z_{i(0)}) \cap \overline{\Delta}(0) \neq \phi$, this is true. If $c(z_{i(0)}) \cap \overline{\Delta}(0) = \phi$, then $c(z_{i(0)})$ must contain $\overline{\Delta}(0)$ in its interior so we will have $\tilde{c}(z_{i(0)}) \supset \overset{\sim}{\Delta}(0)$ (thanks to the inclusion of all the interior bonds that are part of the infinite cluster).

The last construction allows us to associate with each self–avoiding sequence $z_0 = 0, z_1, \ldots z_n = z$, a path from $\overline{\Delta}(0)$ to $\overline{\Delta}(z)$ through the set of open bonds. The next step is to obtain estimates on the length of this path. To do this, we observe that $\tilde{c}(x) \subset \overline{\Delta}(x)$, so (4) implies that $E|\tilde{c}(x)|^m < \infty$ for all $m < \infty$. To obtain estimates on the sum of the lengths, we need to estimate the covariances.

(12) Lemma. For any $q < \infty$ there is a constant $C_q < \infty$ so that

$$|\mathrm{Cov}(|\tilde{c}(x)|,|\tilde{c}(y)|)| \leq C_q \cdot |x-y|^{-q}$$

Proof: Let $D(x,k) = \{ y : |x-y| \leq k \}$. Any circuit that has length k and contains x in its interior must lie in the interior of $D(x,k)$, so if $|x-y| \geq i+j$ the events $\{ |\tilde{c}(x)| = i \}$ and $\{ |\tilde{c}(y)| = j \}$ are independent. Combining this observation with the definition of covariance gives

$$\text{Cov}(|\tilde{c}(x)|, |\tilde{c}(y)|) = \Sigma\ ij\ P(\ |\tilde{c}(x)| = i,\ |\tilde{c}(x)| = j\ )$$

$$- \Sigma\ i\ P(\ |\tilde{c}(x)| = i\ )\ j\ P(\ |\tilde{c}(x)| = j\ )$$

where the sums are over $\{\ (i,j) : i+j > |x-y|\ \}$. If $|x-y| = k$, then each of the sums is smaller in absolute value than

$$2 \underset{i \geq k/2}{\Sigma}\ \overset{i}{\underset{j=1}{\Sigma}}\ i\ j\ P(\ |\tilde{c}(x)| = i\ ) \leq 2 \underset{i \geq k/2}{\Sigma}\ i^3\ P(\ |\tilde{c}(x)| = i\ )$$

$$\leq 2\ (2/k)^q\ E(|\tilde{c}(0)|^{q+3}).$$

To obtain the conclusion stated in (12), combine the estimates on the two sums using the observation: if $a,b \geq 0$ then $|a-b| \leq \max(|a|, |b|)$.

(12) allows us to get an estimate on $L_m = |\tilde{c}(z_1)| + \dots + |\tilde{c}(z_m)|$ which is the analogue of (i) in the proof of (a).

(13) Lemma. Let $z_1, \dots z_m$ be a path with $|z_i - z_j| = |i-j|$ for all i,j. (A sufficient condition for this is $|z_1 - z_m| = m-1$.) There are constants C and $K \in (0,\infty)$ so that

$$P\ (L_m > Km\ ) \leq C/m.$$

Proof: Let $a = E|\tilde{c}(0)|$, $\tau_i = |\tilde{c}(z_i)| - a$, and $S_m = \tau_1 + \dots + \tau_m$.

$$E\ S_m^2 = m\ E\tau_1^2 + 2 \underset{1 \leq i < j \leq m}{\Sigma}\ E\tau_i \tau_j.$$

Using a trivial inequality and then (12) with q = 4, we see that the second term is

$$\leq 2m \sum_{k=2}^{\infty} |E\tau_1 \tau_k| \leq Cm.$$

The last result implies $E\, S_m^2 \leq C'm$. The desired result now follows from the fact that

$$P(\, L_m > (a+1)m \,) = P(\, S_m > m \,) \leq E\, S_m^2/m^2.$$

To prove (b), we need to improve the last result so that the right–hand side is $O(m^{-3})$. To do this, we need to construct three paths from 0 to z that are well separated. For simplicity, we assume $z = (n,0)$. In this case, we let N be the smallest integer $\geq n^{1/4}$, and define the paths to be:

$$r_1 : (0,0) \to (0,2N) \to (n,2N) \to (n,0)$$
$$r_2 : (0,0) \to (n,0)$$
$$r_3 : (0,0) \to (0,-2N) \to (n,-2N) \to (n,0)$$

where again $x \to y$ denotes moving from x to y by a straight line path. Let

$$R_0 = \sum_{j=-2N}^{2N} |\tilde{c}(0,j)|, \qquad R_1 = \sum_{j=-2N}^{2N} |\tilde{c}(n,j)|,$$

and

$$U_h = \sum_{m=1}^{n} |\tilde{c}(m,(4-2h)N)| \quad \text{for } h = 1, 2, \text{ and } 3.$$

From (11) it follows that there is a path from 0 to (n,0) with length at most

$$R_0 + R_1 + \min_{1 \leq h \leq 3} U_h.$$

To estimate R_0 and R_1, we use:

(14) Lemma. If S_m is the sum of m identically distributed random variables X_1, X_2,...
X_m, then

$$ES_m^4 \le 4m^4 EX_1^4.$$

Proof: If we expand S_m^4, we get m^4 terms of the form $EX_i X_j X_k X_\ell.$ These terms are each

$$\le E(\max(X_i,X_j,X_k,X_\ell))^4 \le \int_0^\infty 4x^3 \, 4 \, P(X_1 > x) \, dx = 4 \, EX_1^4.$$

Applying Chebyshev's inequality, and using (14) with $X_i = |\tilde{c}(0,-2N + i - 1)|$ gives

$$P(S_m > m^\beta) \le ES_m^4/m^{4\beta} \le 4E|\tilde{c}(0,0)|^4/m^{4(\beta-1)}.$$

Setting $\beta = 4$ and $m = 4N + 1 \le 4n^{1/4} + 5$ gives

(15) $$P(R_0 > Kn) \le C/n^3.$$

To estimate min U_h, we need to show that the circuits associated with these paths are almost independent. To do this, we break the plane into three regions: $D_1 = \{ x : x_2 \ge N \}$, $D_2 = \{ x : N > x_2 > -N \}$, and $D_3 = \{ x : x_2 \le -N \}$. The probability that some circuit on one of the paths $(1,(4-2h)N) \to (n,(4-2h)N)$ lies outside its region is no more than

$$3n \, P(|\tilde{c}(0)| \ge N) \le 3E|\tilde{c}(0)|^{16} n^{-3},$$

by Chebyshev's inequality (recall $N \geq n^{1/4}$).

To compare the U_h with independent random variables, we will use a special construction. Generate by independent mechanisms three sets of passage times for the whole plane, and let $\overline{U}_h$ be the total lengths of the circuits associated with the three paths in their respective planes. By taking region D_h from plane h, we can define a fourth plane. If we let Ω' be the event that all the circuits in the definition of of the $\overline{U}_h$ lie in their respective regions, then on Ω', $U_h = \overline{U}_h$ for h = 1,2,3. From this observation it follows that

$$P(\min U_h > Kn, \Omega') = P(\min \overline{U}_h > Kn, \Omega') \leq P(U_1 > Kn)^3.$$

Applying (13) (recall U_h and L_n have the same distribution) and results in the last two paragraphs, shows

$$P(\min U_h > Kn) \leq Cn^{-3}.$$

Combining the last result with (15) gives

(16) $$P(\hat{t}(0, ne_1) > Kn) \leq Cn^{-3},$$

the analogue in this setting of (ii) from the proof of (a). The last bound can be summed over n to prove the analogue of (iii). Then, as explained in the proof of (a), the Denouement of Chapter 1 takes over, and proves that there is a convex set A so that for any $\epsilon > 0$

$$P((1-\epsilon)tA \subset \hat{\xi}_t \subset (1+\epsilon)tA) \to 1 \text{ as } t \to \infty.$$

Since $\hat{\xi}_t \supset \bar{\xi}_t$, the last result implies the first conclusion in (b).

To prove the second statement we observe that for any $\epsilon > 0$

$$|t^{-1}\bar{\xi}_t \, \triangle \, \{x : \mu(x) \le 1\}| \le |\{x : \mu(x) \le 1-\epsilon\} - t^{-1}\bar{\xi}_t|$$
$$+ |\{x : (1-\epsilon) < \mu(x) \le (1+\epsilon)\}|$$
$$+ |t^{-1}\bar{\xi}_t \cap \{x : \mu(x) > (1+\epsilon)\}|.$$

We have proved that the third term (on the right hand side) is eventually 0. The second, by scaling, equals $((1+\epsilon)^2 - (1-\epsilon)^2)|\{x : \mu(x) \le 1\}|$, which is small if ϵ is small. So it suffices to show that the first term $\to 0$ for any $\epsilon > 0$. Since $t(0,z) \le u(0) + \hat{t}(0,z) + u(z)$,

$$\{x : \mu(x) \le (1-\epsilon)\} - t^{-1}\bar{\xi}_t \, \subset \, \{x : \mu(x) \le (1-\epsilon), \, u(0) + \hat{t}(0,tx) + u(tx) > t\}$$

(Here we are supposing that $u(y)$ has been extended to $y \in \mathbb{R}^2$ in the usual way.) It follows from the limit theorem for $\hat{\xi}_t$ that if $\delta < \epsilon$ then

$$\{x : \mu(x) \le (1-\epsilon), \, u(0) + \hat{t}(0,tx) > (1-\delta)t\} = \phi$$

for t large (with probability 1), so it is enough to show that

$$|\{x : \mu(x) \le 1 - \epsilon, \, u(tx) \ge \delta t\}| \to 0.$$

If we pick K large enough so that $\{x : \mu(x) \le 1-\epsilon\} \subset [-K,K]^2$ (here we use $\mu(x) > 0$ which comes from our assumption that $F(x) > 0$), then

$$|\{\, x : \mu(x) \le 1-\epsilon,\, u(tx) \ge \delta t \,\}| \le t^{-2} \sum_{z\in[-Kt,Kt]^2} 1(u(z) > \delta t).$$

If we replace δt by M in the last expression, then the multiparameter ergodic theorem (see Dunford (1951), or for a subadditive version Smythe (1976)), implies that the limit is $K^2 P(u(z) > M)$. Letting $t \to \infty$, and then $M \to \infty$, we see that the unwanted term approaches 0, and the proof is complete.

In the next section, we will see that $\mu(x)\equiv 0$ if (and only if) $F(0) \ge 1/2$ so the limiting shape $A = \mathbb{R}^2$. In this case, we cannot have $\bar{\xi}_t \supset (1-\epsilon)tA$, so the result has to be reformulated. In the case $E(Y^2) < \infty$, the conclusion becomes: for all compact K

$$P(\,\bar{\xi}_t \supset tK\,) \to 1 \text{ as } t \to \infty.$$

We leave it to the reader to reformulate (b). The answer, and proofs for a general F, can be found in Cox and Durrett (1981).

8b Critical Behavior of the Bernoulli Model

In this section, we investigate what happens when the passage time distribution in first passage percolation is the Bernoulli distribution, that is, $P(X(e) = 0) = p$ and $P(X(e) = 1) = 1-p$. If we let $t_{0,n} = t(0, ne_1)$ to simplify our notation, then (1) in the last section implies that there is a constant $\mu(p)$ so that as $n \to \infty$

$$t_{0,n}/n \to \mu(p) \text{ a.s.}$$

The first result identifies $\mu(p)$ for $p > 1/2$.

(1) Theorem. If $p > 1/2$, then $\mu(p) = 0$.

Proof: If we call bonds with passage time 0 open, and those with passage time 1 closed, then for $p > 1/2$ there is positive probability that C_0, the cluster of open bonds containing the origin, is infinite. I claim that infinitely many points on the x axis are part of the infinite cluster of open bonds (which was proved to be unique in Section 6a). If you accept this claim, then the result follows, because it implies that on $\{ |C_0| = \infty \}$ we have $t_{0,n} = 0$ for infinitely many n, and hence the constant which is the almost sure limit of $t_{0,n}/n$ must be 0.

To prove the claim, observe that the ergodic theorem implies

$$\frac{1}{n} \sum_{m=1}^{n} 1(|C(m,0)| = \infty) \to W$$

where $C(m,0)$ is the cluster of open bonds containing $(m,0)$, and W is a random variable with $EW = P(|C_0| = \infty)$. W is measurable with respect to the state of the bonds in $\mathbb{R}^2 - [-N,N]^2$ for any N, so the 0–1 law* implies that W is constant, and it follows that $\{ m \geq 1 : |C(m,0)| = \infty \}$ is infinite with probability 1.

        *Note: For more details on how the 0–1 law is applied, see the proof of (12) in Section 6a.

Exercise (due to Ted Cox). Let $Z(x)$ be the travel time from x to the infinite cluster of bonds with passage time 0. Clearly, $t_{0,n} \leq Z(0,0) + Z(n,0)$. Show that as $n \to \infty$, $t_{0,n}$ converges in distribution to $Z_1 + Z_2$ where the Z_i are independent and have the same distribution as $Z(0,0)$.

For the rest of this section, we are concerned with what happens when $p \leq 1/2$. The first result in that case is:

(2) Theorem. If $p < 1/2$ then $\mu(p) > 0$.

Proof: To prepare for the real proof, we will warm up by proving that if $p < 1/3$, then $\mu(p) > 0$. To do this, we observe that if $t_{0,n} \leq \delta n$, then there is a self–avoiding path of length n with travel time $\leq \delta n$. To estimate the probability of such a path, we observe that if $\theta > 0$ and $S_n = X_1 + ... + X_n$, where the X_i are independent with $P(X_i = 0) = p$ and $P(X_i = 1) = 1-p$, then

$$e^{-\theta \delta n} P(S_n \leq \delta n) \leq E\exp(-\theta S_n) = (p + (1-p)e^{-\theta})^n.$$

If $p < 1/3$, then we can pick $r > 1$ so that $r^2 p < 1/3$, pick θ large so that $p + (1-p)e^{-\theta}$

$\le$ rp, and then δ small so that $e^{\theta\delta} \le$ r. If we do this, then the estimate above implies

$$P(\, S_n \le \delta n \,) \le (e^{\theta\delta}(p + (1-p)e^{-\theta}))^n \le (r^2 p)^n.$$

The number of self–avoiding paths of length n is $\le 4 \cdot 3^{n-1}$ so

$$P(\, t_{0,n} \le \delta n \,) \le 4 \cdot 3^{n-1}(r^2 p)^n \to 0 \ \text{as } n \to \infty.$$

To reduce p < 1/2 to the case treated above, we will use a "block construction." We introduce a "renormalized lattice" $\mathscr{L} = \{\, (2Mm, 2Mn) : (m,n) \in \mathbb{Z}^2 \,\}$ where M = inf $\{\, L : 4\rho_{6L,2L} < \epsilon_1 \,\}$, and ϵ_1 is a small constant which will be chosen at the end of the proof. Let A_M = the boundary of $[-M,M]^2$, and B_M = the boundary of $[-3M,3M]^2$. We call bonds with passage time 0 open and those with passage time 1 closed. If $\alpha \in \mathscr{L}$, we say α is open if there is an open path from $\alpha + A_M$ to $\alpha + B_M$. To explain the definition of M, observe that if there is an open path from A_M to B_M, then there is an open crossing of $(M,3M) \times (3M,-3M)$, or of one of the other three rectangles obtained by rotating this one by 90, 180, or 270 degrees. (See Figure 8.3.)

To prove (2), we will (i) show that if there is a path from 0 to (n,0) with travel time $\le$ n/10M then there is a self–avoiding path on $\mathscr{L}$ that has length at least n/4M and in which at least 1/2 of the sites are open, and then (ii) show that the probability of such a path $\to$ 0 as n $\to \infty$. To get a path on $\mathscr{L}$ from a self–avoiding path $z_0 = 0, z_1,$ ... $z_k =(n,0)$, let $\alpha_0 = 0$, and let $m_0 = \inf \{\, m : z_m \notin \alpha_0 + (-3M,3M)^2 \,\}$. $z(m_0)$ is in the boundary of one, or possibly two, of the squares $\alpha_0 + 2Mv + [-M,M]^2$ where

$$v \in \mathscr{N} = \{\, (1,2),\ (0,2),\ (-1,2),\quad (2,1),\ (2,0),\ (2,-1),$$
$$(1,-2),(0,-2),(-1,-2),\quad (-2,1),(-2,0),(-2,-1) \,\}$$

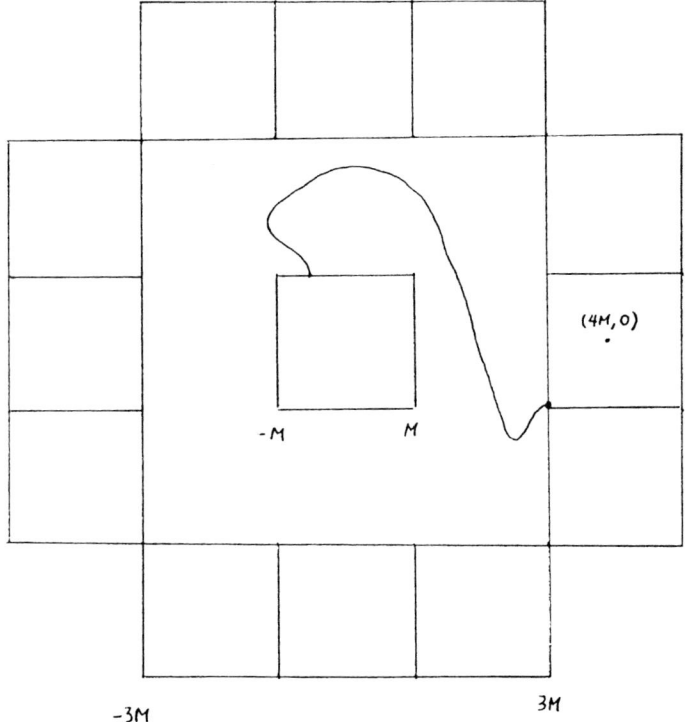

Figure 8.3

(Observe that $z(m_0)$ cannot be (3M,3M), or one of the other three corners.) If the square containing $z(m_0)$ is unique, let α_1 be the center of that square. If $z(m_0)$ is in two squares, then choose the one that is higher or to the right.

　　With the first step described, it should be clear how to repeat the construction. Let $m_i = \inf \{ m > m_{i-1} : z_m \notin \alpha_{i-1} + (-3M,3M)^2 \}$ with $m_i = \infty$ if the set is ϕ. If $m_i < \infty$, let $\alpha_i =$ the center of the square on whose boundary $z(m_i)$ lies (breaking ties as before). Let $I = \sup \{ i : m_i < \infty \}$. $\alpha_0, \alpha_1, \ldots \alpha_I$ may not be self–avoiding (see Figure 8.4), but by deleting loops we can make a self–avoiding path $\beta_0 = 0, \beta_1 \ldots \beta_J$.

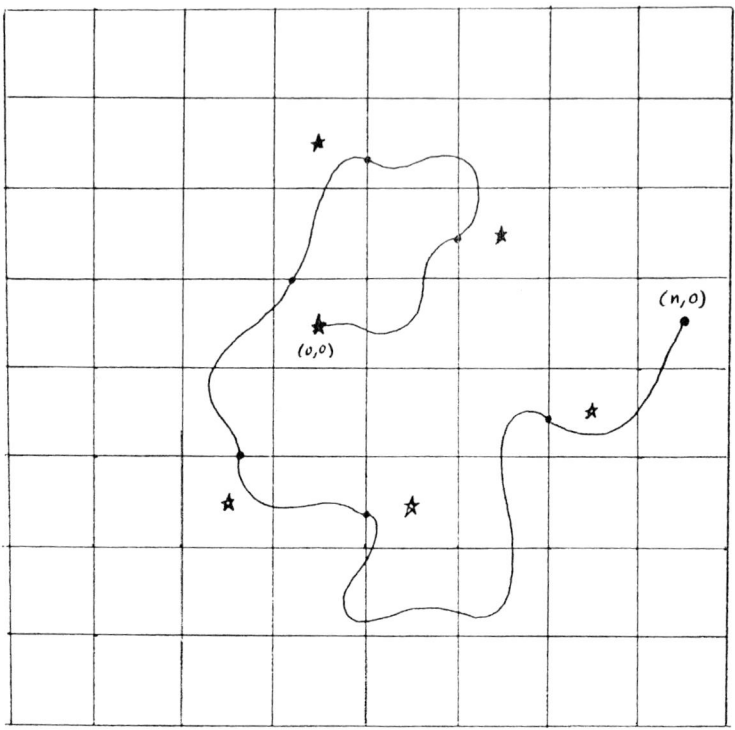

Figure 8.4

To accomplish (ii), we begin by observing that $\beta_j - \beta_{j-1} \in \mathcal{N}$ for $1 \leq j \leq J$, and $(n,0) \in \beta_J + (-3M, 3M)^2$, so $J \geq [n/4M] - 1$ where $[x]$ = the largest integer $\leq x$. If the original path has travel time $\leq n/10M$, and n is large, then at least $1/2$ of the β_j must be open. The events $\{ \beta_j$ is open $\}$ are not independent, but if $\| \beta_i - \beta_j \|_\infty \geq 6M$ then they are. For a given $\alpha \in \mathcal{L}$ there are 25 $\beta \in \mathcal{L}$ with $\| \alpha - \beta \|_\infty < 6M$. Hence, we can find $\{ \gamma_1, \cdots \gamma_K \} \subset \{ \beta_0, \cdots \beta_J \}$ so that all the γ_k are open, $\| \gamma_i - \gamma_j \|_\infty \geq 6M$ when $i \neq j$, and $K = \lceil J/50 \rceil$ = the smallest integer $\geq J/50$.

Combining the last two paragraphs, observing that there are at most 12^J paths

$\beta_0{=}0,\ \beta_1\ \dots\ \beta_J$, and letting $J_0 = [n/4M] - 1$, we get

$$P(\ t_{0,n} \le n/10M\ ) \le \sum_{j=J_0}^{\infty} 12^j \left[\begin{array}{c} j \\ \lceil j/50 \rceil \end{array} \right] (\epsilon_1)^{\lceil j/50 \rceil}$$

$$\le \sum_{j=J_0}^{\infty} 24^j\ \epsilon_1^{\ j/50},$$

where in the last inequality we have used the fact that the binomial coefficient is $\le 2^j$. If we pick $\epsilon_1 = (1/48)^{50}$ then the last sum $\to 0$ as $n \to \infty$, and we have proved (2).

Having worked to prove (2), the reader should be happy to hear that we have proved more than we have advertised. If we call bonds with passage time 0 open, those with passage time 1 closed, and let C_0 be the cluster of open bonds containing 0, then:

(3) Corollary. There are constants $C, \gamma \in (0,\infty)$ (independent of p) so that if $p < 1/2$ then

$$P(\ (n,0) \in C_0\ ) \le Ce^{-\gamma n/M(p)}$$

where $M(p) = \inf \{\ L : 4\rho_{6L,2L} < (1/48)^{50}\ \}$.

Proof: $(n,0) \in C_0$ if and only if $t_{0,n} = 0$, so summing the series at the end of the proof of (2) gives

$$P(\ (n,0) \in C_0\ ) \le 2^{-[n/4M(p)]+2}.$$

(4) Corollary. Let $L(p) = \inf \{\ p : \rho_{L,L} < \epsilon_0\ \}$, where ϵ_0 is the magic constant of

Section 6a. There is a constant c > 0 (indpendent of p) so that if p < 1/2 then

$$\mu(p) \geq c/L(p).$$

Proof: The proof of (2) shows that $\mu(p) \geq 1/10M(p)$. (10) and (5) from Section 6a, applied to closed crossings of the dual sponge, show there is a constant C (which only depends on ϵ_1 and hence is independent of p) so that $M(p) \leq C\, L(p)$.

At this point the reader may be doubting our sanity for trying to extract explicit bounds on $\mu(p)$ from the proof of (2). The next result should explain this and why we are interested in constants independent of p.

(5) Theorem. Let $K(p) = \sup \{ L : \rho_{L,L} \geq \epsilon_0 \}$ There is a constant $C < \infty$ (independent of p) so that if p < 1/2 then

$$\mu(p) \leq C/K(p).$$

Remark. The reader should note that $K(p) \geq L(p) - 1$, and (4) and (5) combine to show $K(p) \leq (C/c)\, L(p)$, so $\mu(p) \approx 1/L(p)$ as $p \uparrow 1/2$. We will have more to say about this later, but for the moment the reader should note that $L(p) \to \infty$ as $p \uparrow 1/2$ so (5) implies $\mu(1/2) = 0$, and when combined with (4), describes the rate at which $\mu(p) \to 0$.

Proof: Consider the first passage problem is the region $\{ z \in \mathbb{Z}^2 : |z_2| \leq T/2 \}$ which we will call the "time tunnel" of width T. Let $t_{0,n}^{T}$ denote the passage time between the origin and (n,0) along bonds that lie in the tunnel, and let $s_{0,n}^{T}$ denote the passage time in the tunnel from the line $z_1 = 0$ to the line $z_1 = n$. The next result connects these two new passage times with $t_{0,n}$.

(6) Lemma. As $n \to \infty$

$$s_{0,n}^T/n \text{ and } t_{0,n}^T/n \to \mu^T(p) \text{ a.s.,}$$

and as $T \to \infty$

$$\mu^T(p) \downarrow \mu(p).$$

Proof: If we define $s_{m,n}^T$ in the obvious way then it is clear (recall all bonds have passage time ≤ 1) that

$$s_{0,m}^T + T + s_{m,n}^T \geq s_{0,n}^T.$$

So the subadditive ergodic theorem can be applied to $u_{m,n} \equiv s_{m,n}^T + T$ to conclude

$$u_{0,n}/n \to \mu^T(p) \text{ a.s.}$$

This gives the result for $s_{0,n}^T/n$. To get the result for $t_{0,n}^T/n$ observe that

$$s_{0,n}^T \leq t_{0,n}^T \leq s_{0,n}^T + 2T.$$

To prove that $\mu^T(p) \downarrow \mu(p)$, observe that $\mu^T(p) \geq \mu(p)$, and the bounded convergence theorem implies

$$(n \geq) \; Et_{0,n}^T \downarrow Et_{0,n} \text{ as } T \uparrow \infty.$$

The last result and the fact that $Et_{0,n}/n \to \mu(p)$ as $n \to \infty$ imply that if we pick n large and then T large

$$\mu(p) + 2\epsilon \geq Et_{0,n}/n + \epsilon \geq Et_{0,n}^T/n \geq \mu^T(p),$$

the last inequality following from the fact that $t_{0,n}^T$ is subadditive.

To prove (5), we will look at the end to end tunnel times $s_{0,n}^T$ because these times have a geometric interpretation. Let us associate with each unit time bond a "barrier" on the dual lattice $Y^2 = (1/2,1/2) + \mathbb{Z}^2$, where the barrier is the dual bond that crosses it. Then, a little graph theory tells us that $s_{0,n}^T$ is the number of disjoint "barrier surfaces" that separate $\{ z_1 = 0 \}$ from $\{ z_1 = n \}$ in the time tunnel, where a barrier surface is a self–avoiding path of barriers on the dual that connects $z_2 = -T/2$ to $z_2 = T/2$.

To estimate $s_{0,n}^T$ when $T = K(p)$ and $n = NK(p)$, we will call $\{ z : (j–1)K(p) \leq z_1 < jK(p), |z_2| \leq K(p)/2 \}$ the jth segment of the time tunnel, and let Q_j be the number of disjoint barrier surfaces in $\{ z : |z_2| \leq T/2 \}$ that start from some point on the bottom of the jth segment. The Q_j form a stationary sequence (since we have discarded the requirement that the barriers stay in $\{ z : 0 \leq z_1 \leq NK(p) \}$), and have

$$t_{0,NK(p)}^{K(p)} \leq \sum_{k=1}^{N} Q_j.$$

Dividing both sides of the last equation by N and letting $N \to \infty$ gives

$$K(p) \, \mu^{K(p)} \leq E(Q_1).$$

(The limit is constant because again, it is measurable with respect to the tail σ–field of an i.i.d. sequence.) The last detail, then, is to give a bound on $E(Q_1)$ that is independent of p.

To bound $E(Q_1)$ we begin by observing that

$$P(\, Q_1 = 0 \,) \geq (\rho_{K,K})^2 \cdot \rho_{3K,K}$$

(where K is short for K(p)), because if there is a left to right crossing of $[-K,2K] \times [-K/2,L/2]$, and top to bottom crossings of $[-K,0] \times [-K/2,K/2]$ and $[K,2K] \times [-K/2,K/2]$ by zero time bonds, then $Q_1 = 0$. (See Figure 8.5.) From the definition of

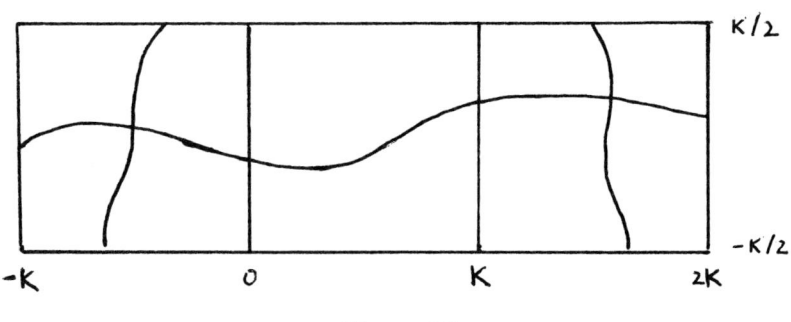

Figure 8.5

K, it follows that $\rho_{K,K} \geq \epsilon_0$. Applying (2) and (3) from Section 6a, gives

$$\rho_{3K,K} \geq (1 - (1-\epsilon_0)^{1/2})^{27},$$

so

$$P(\, Q_1 = 0 \,) \geq \delta > 0,$$

where δ depends only on ϵ_0, and hence is independent of p.

To estimate the rest of the distribution we will show

(7) $P(Q_1 \geq k+1 \mid Q_1 \geq k) \leq (1-\delta)$ for $k \geq 1$.

To prove this we begin with the case $k = 1$. Let s be the leftmost barrier surface that starts at the bottom of the first segment. s divides $\{ z : |z_2| < K/2 \}$ into two connected components. We will call these two components the sets of points that lie to the left and right of s respectively. The event that the leftmost barrier is s, is measurable with respect to the bonds that lie on s and to the left of it; and, hence, if we condition on the value of s the bonds that lie to the right are independently open and closed with probabilities p and $(1-p)$.

Given that the leftmost barrier is s, the probability that we can find a second barrier is the probability that we can find a barrier that starts at the bottom of the first segment and lies in the set of bonds lying to the right of s. This is harder than finding a barrier without the second restriction, so (7) holds when $k = 1$. The proof for $k > 1$ is almost the same and is left to the reader. With (7) established, we are done since

$$E(Q_1) = \sum_{k=1}^{\infty} P(Q_1 \geq k) \leq 1/\delta.$$

Results $(2) - (5)$ should make it clear that $L(p)$ is an important length scale for percolation and first passage percolation. In the physics literature $L(p)$, though not defined in exactly this way, is called the "correlation length." A comprehensive discussion of this concept and its many formulations could itself fill a book. Roughly speaking, if one lets p $\uparrow$ 1/2 and looks at the clusters in the percolation model, then one should scale space by the correlation length to get an interesting limit.

For a discussion of the correlation length and its relationship to scaling theory, see Kesten (1987d). We will content ourselves here to (i) tease the reader by saying that physicists tell us that $L(p) \approx (1/2-p)^{-4/3}$ as $p \uparrow 1/2$ where $\approx$ means that the ratio of the logarithms of the two quantities $\to 1$, and (ii) prove a rigorous lower bound on $L(p)$ (due to J.T. Chayes, L. Chayes, D. Fisher, and T. Spencer(1986)) which is three–fourths as good.

(8) Theorem. $\liminf_{p \uparrow 1/2} L(p)/(1/2-p) > 0.$

Proof: The key to the proof is the following ridiculously simple inequality:

(9) Lemma. Consider independent bonds that are open with probability p. If A is any event depending upon the state of N bonds, then

$$\frac{d}{dp} P_p(A) \le (N/p(1-p))^{1/2}$$

Proof: Supposing A depends upon the bonds in S, and letting $\omega \in \{0,1\}^S$ denote a possible outcome for the bonds in S (0 = open, 1 = closed), we have

$$P_p(A) = \Sigma \, P_p(\omega) \, 1_A(\omega),$$

where $1_A(\omega) = 1$ if A happens in the realization ω and 0 otherwise. Since

$$P_p(\omega) = p^{n(\omega)}(1-p)^{N-n(\omega)},$$

where $n(\omega) =$ the number of 0's in ω, it is clear that

$$\frac{d}{dp} P_p(\omega) = \left(\frac{n(\omega)}{p} - \frac{N-n(\omega)}{1-p} \right) P_p(\omega) = \frac{(n(\omega)-Np)}{p(1-p)} P_p(\omega).$$

Summing over ω gives

$$\left| \frac{d}{dp} P_p(A) \right| \leq (p(1-p))^{-1} E_p |n(\omega)-Np|$$

$$\leq (p(1-p))^{-1} (E_p |n(\omega)-Np|^2)^{1/2}$$

$$= (N/p(1-p))^{1/2},$$

since the variance of the sum of N independent Bernoulli random variables is $Np(1-p)$.

With (9) in hand, the rest is easy. Letting A = {there is a left to right crossing of (0,L) x (0,L) } and using (9), shows that there is a constant $c > 0$ so that if $p \geq 1/2 - c/L$ then $\rho_{L,L} > \epsilon_0$. The last result implies $K(1/2 - c/L) \geq L$, and the desired result follows from remarks after (5).

Our last topic in this section is to consider the behavior of $t_{0,n}$ when $p = 1/2$.

(10) Theorem. There are constants c,C $\in (0,\infty)$ so that

$$c \log n \leq E_{1/2}(t_{0,n}) \leq C (1 + \log n).$$

The reader should note the 1 on the right is needed only for the case $n = 1$.

Proof: To prove the lower bound, we let T_n be time it takes the fluid to escape from the box $B_n = (-n,n)^2$. If we define barriers to be bonds on the dual lattice Y^2 that cross bonds with travel time 1, then it follows from results in Section 6a that the probability of a closed circuit in the annulus $(-3m,3m)^2 - (m,m)^2$ is bounded below by $\delta > 0$. If $n \geq 3^N$, then we can put N disjoint annuli $A_j = (-3^j,3^j)^2 - (-3^{j-1},3^{j-1})^2$, $1 \leq j \leq N$, in $(-n,n)^2$, so

$$Et_{0,n} \geq ET_n \geq \delta N,$$

which proves the lower bound. (Here and in what follows, we will drop the subscript 1/2 from the expected value.)

To prove the upper bound, we will be begin by deriving an upper bound for $S_N = T(3^N)$. Geometrically speaking, we are trying to estimate the number of disjoint barrier rings that separate 0 from the outside of $(-3^N,3^N)^2$. Any such ring may visit several of the annuli A_1, A_2, ... A_N, so we start a classification scheme for rings based on the "outer reach" of the ring = the largest of our annuli visited by the ring, and denote by V_k the number of disjoint barrier rings whose outer reach is k. Clearly,

$$S_N \leq \sum_{k=1}^{N} V_k,$$

so what we need is bound on $E(V_k)$ which is independent of k.

To this end, we define in an analogous fashion the inner reach of a ring, and let V_k^j, $0 \leq j \leq k$ be the number of independent rings with outer reach k and inner reach k−j. It should be clear that

$$V_k \leq \sum_{j=0}^{k} V_k^j.$$

To prove the desired bound we will show

(11) **Lemma.** $E(V_k^j) \leq C r^j$ where $r < 1$.

Proof: We begin with the case $j = 0$. V_k^0 = the number of barrier rings that stay in the annulus A_k. It is easy to see that if there is a crossing of $(3^{k-1}, 3^k) \times (-3^{k-1}, 3^{k-1})$ by open bonds on $\mathbb{Z}^2$ there is no barrier ring in A_k on the dual, so $P(V_k^0 = 0) \geq \delta > 0$ independent of k. By locating the innermost barrier ring, and repeating the argument we used in the proof of (5) to estimate $E(Q_1)$, it follows that $P(V_k^0 \geq n) \leq (1-\delta)^n$, and summing over $n \geq 1$ gives $E(V_k^0) \leq (1-\delta)/\delta$.

A similar argument covers the case $j = 1$, so we turn now to $j \geq 2$. In order to form a barrier surface with outer reach k and inner reach k−j, it has to be the case that none of the intervening j−1 annuli contain a complete circuit of open bonds. The probability of such a circuit is bounded below by $\eta > 0$, so

$$P(V_k^j \geq 1) \leq (1-\eta)^{j-1},$$

and repeating the argument in the last paragraph shows

$$P(V_k^j \geq n) \leq ((1-\eta)^{j-1})^n.$$

Summing over $n \geq 1$ gives

$$E(V_k^j) \leq (1-\eta)^{j-1}/\eta,$$

the estimate that we promised to prove. (Here we have replaced the $(1-(1-\eta)^{j-1})$

which should be in the denominator by η, its value when $j = 2$).

Using (11) and adding everything up gives

$$E(V_k) \leq \sum_{j=0}^{k} E(V_k^j) \leq C/(1-r),$$

so if $n \leq 3^{N+1}$ we have

$$ET_n \leq ES_{N+1} \leq (N+1)C/(1-r).$$

To obtain the analogous result for $t_{0,n}$, we begin by observing that $t_{0,n}$ is the number of barrier rings that separate the origin from $(n,0)$. These rings fall into one of two categories: (i) those which surround 0 but do not enclose $(n,0)$, and (ii) those which surround $(n,0)$ but do not enclose 0. The expected number of rings of each type is equal, so we will go to work on those of type (i) and double the bound when we are done.

In the argument above, we have counted the number of barrier rings with outer reach $\leq N+1$. Using (11) we can estimate the number of barrier rings with outer reach $> N+1$ that do not enclose $(n,0)$ by observing that such a ring must touch A_{N+1}, so the expected number is at most

$$\sum_{k=2}^{\infty} \sum_{j=k-1}^{\infty} E(V_{N+k}^j) \leq \sum_{k=2}^{\infty} \sum_{j=k-1}^{\infty} C\,r^j$$

$$= \sum_{k=2}^{\infty} C\,r^{k-1}/(1-r) \leq Cr/(1-r)^2$$

a constant that is independent of n.

The results in this section are from Chayes, Chayes, and Durrett (1988). (2) and (4) are improvements due to H. Kesten. To close the section, we have two simulations of first passage percolation with p = 1/2. The first shows the points with passage time 0, and the second the points with passage time ≥ 2.

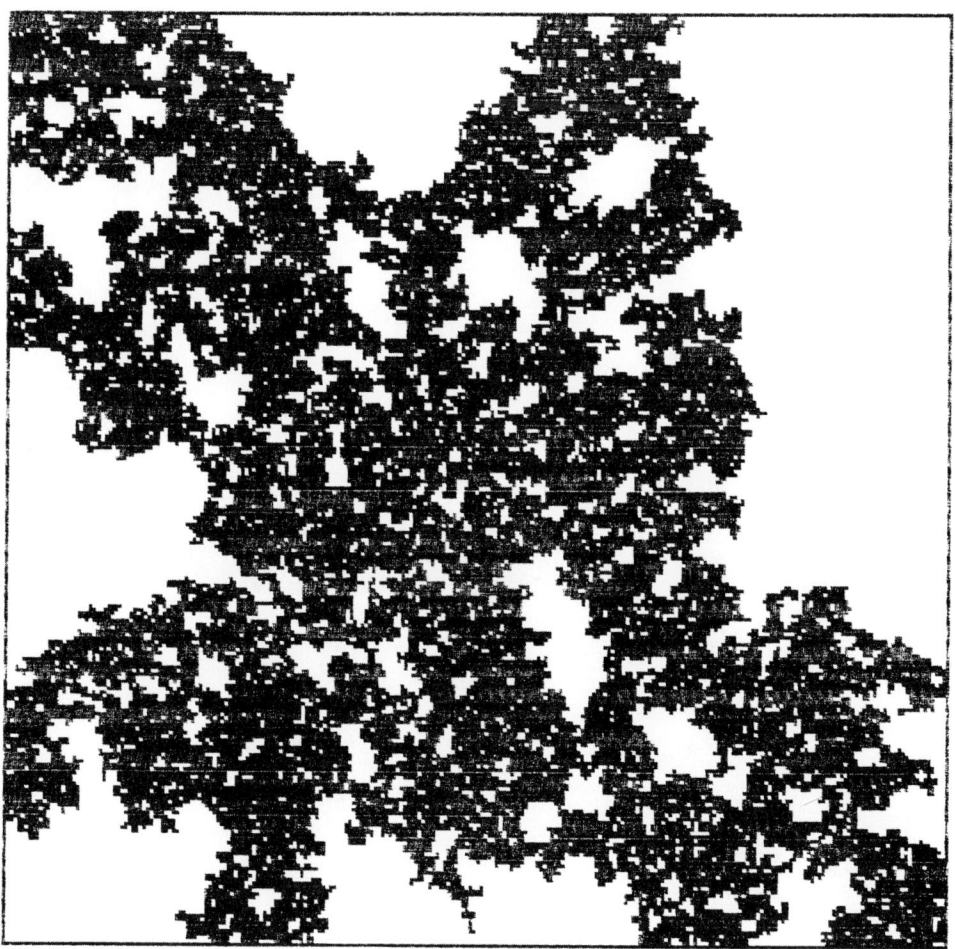

Figure 8.6 Points with passage time 0.

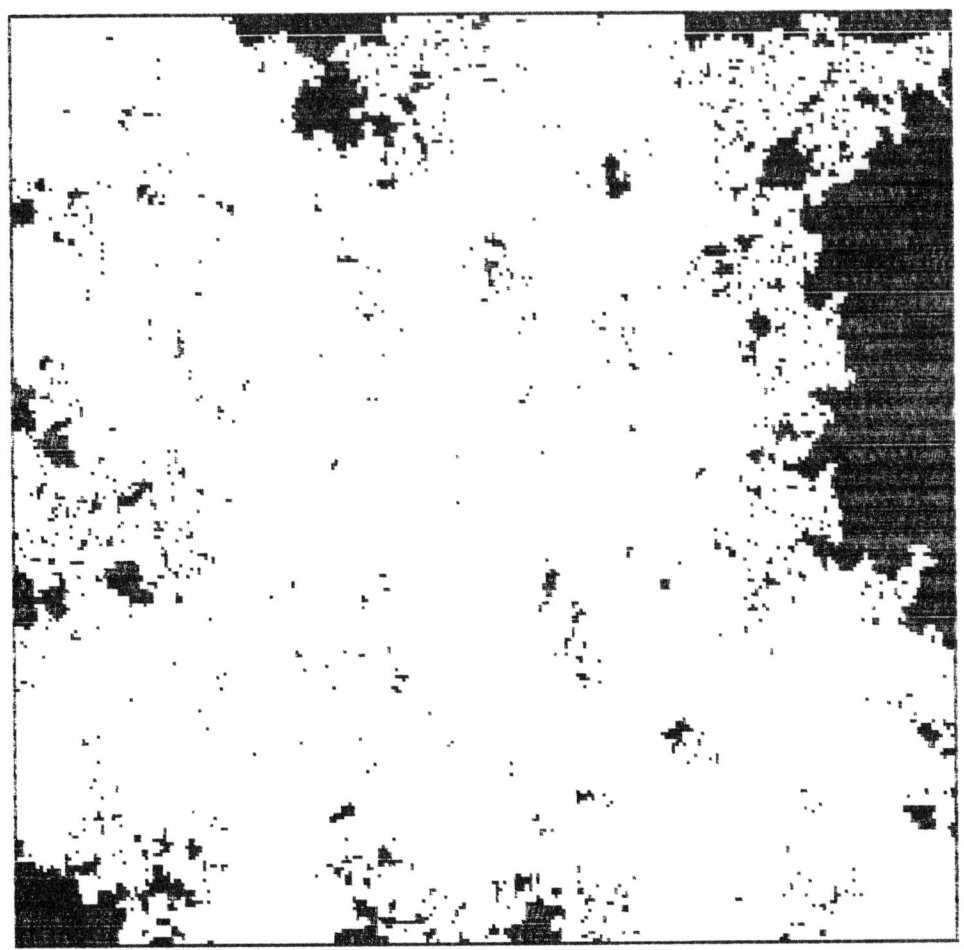

Figure 8.7   Points with passage time ≥ 2.

9 Epidemic Models

In this chapter, we prove a result that describes the asymptotic behavior of a process which has been used to model the spread of an epidemic or a forest fire. In this model, each site $x \in Z^2$ can be in one of three states: 1, i, or 0; and the state of the process is represented by a function $\xi_t : Z^2 \to \{1,i,0\}$, $\xi_t(x)$ giving the state of x at time t. In the epidemic interpretation, 1 = healthy, i = infected, 0 = immune; while for a forest fire, 1 = a live tree, i = on fire, and 0 = burnt. In what follows, we will start out using the epidemic formulation of the model, and half way through switch over to talking about forest fires. We begin by describing the model in epidemic language, following Mollison (1977).

An infected individual emits germs according to a Poisson process with rate α. A germ emitted from x goes to one of its four neighbors x+(1,0), x+(0,1), x+(−1,0), x+(0,−1), choosing from these possibilities with equal probability. If the germ goes to the site of a healthy individual, then that individual immediately becomes infected and begins to emit germs. She stays infected for a random amount of time with distribution F, then recovers and is immune from further infection. To complete the description, we declare that the infection periods and Poisson processes of germs associated with different sites are independent.

In formulating the dynamics above, we are thinking of a disease like measles, where recovered individuals cannot get the disease again. Given the description in the last paragraph, the reader could undoubtedly construct the process but for what follows, it will be useful to have a special construction which we will give now. Let T_x, $x \in Z^2$ be independent random variables with distribution F, and for x,y $\in Z^2$ with

$|x-y| = 1$, let $e(x,y)$ be independent random variables with $P(\ e(x,y) > t\ ) = \exp(-t\alpha/4)$. T_x gives the amount of time x will stay infected (if it ever becomes infected), and $e(x,y)$ is the time lag from the infection of x until the first germ from x is sent to y. With the last interpretations in mind, we let $X(x,y) = e(x,y)$ if $T_x > e(x,y)$, $X(x,y) = \infty$ if $T_x \leq e(x,y)$. We say the oriented bond (x,y) from x to y is open in the first case, and closed in the second. Given the definitions of T_x and $e(x,y)$, it should be clear that the bond (x,y) is open if x tries to infect y during the one time period it is infected, and $X(x,y)$ gives the time lag from the infection of x until it tries to infect y, with $X(x,y) = \infty$ if this never happens.

Imitating previous developments, we make the following definitions:

$x \rightarrow y$ (y can be reached from x) if there is an open path from x to y; that is there is a sequence $x_0 = x, \ldots x_n = y$ of points in $\mathbb{Z}^2$ so that for each $m \leq n$ the oriented bond from x_{m-1} to x_m is open.

The cluster containing 0, $C_0 = \{\ x : 0 \rightarrow x\ \}$.

The relationship of C_0 to the epidemic model is explained by:

(1) Lemma. C_0 = the set of sites that will ever be infected when initially the origin is infected and all the other sites are healthy.

Proof: Clearly if y becomes infected at some time then $y \in C_0$ (because y was infected by a neighbor, who was infected by a neighbor, ... who was infected by 0). We argue the other inclusion by induction on the length of the shortest path to x. If the path has length 1, this is clear, because when 0 tries to infect it either (a) the site is not yet infected and it becomes infected, or (b) the site is already infected, and in either case the site becomes infected. If $x \in C_0$, and $x_0 = 0, x_1, \ldots x_n = x$ is a shortest path to x, then we can apply the last argument with 0 replaced by x_{n-1}. Since the shortest path to x_{n-1} has length $n-1$, the induction hypothesis implies x_{n-1} will become infected. When x_{n-1} tries to infect x_n either (a) or (b) above will happen, and in either case, we conclude that x_n will become infected.

The realtionship of the epidemic model to the percolation process described above was first noticed by Mollison (1977) (see p.322), and developed by Kuulasmaa (1982) who called the resulting structures "locally dependent random graphs," a phrase that means that bonds that begin at different sites are independent. By using ideas from the proof of the clutter percolation theorem of McDiarmid (1980), he was able to prove a useful comparison theorem. To state this result we need some notation: if $x \in \mathbb{Z}^2$ and $A \subset \{ y : |x-y| = 1 \}$, let $\varphi(x,A) = P(\text{all the bonds } (x,y) \ y \in A \text{ are closed})$. φ is called the "zero function" of the epidemic model.

(2) Comparison Theorem. Let B be a collection of paths (each of which has finite length) in $\mathbb{Z}^2$, and let $\mathscr{B}$ be the event that some path in B is open. If two epidemic models have $\varphi_1(x,A) \geq \varphi_2(x,A)$ for all A then $P_1(\mathscr{B}) \leq P_2(\mathscr{B})$.

Proof: We will give what we think is a complete proof. The reader can find more details in Kuulasmaa (1982), pages 749–750. The fact above is true for general locally dependent random graphs, so we prove it first for finite graphs and then we take limits to get the result for $\mathbb{Z}^2$.

To do the first step, we suppose in addition that there is only place where the two zero functions differ; that is, there is an x_0 so that if $x \neq x_0$, then $\varphi_1(x,A) =$

$\varphi_2(x,A)$. To prove the result in this case, we condition on the state of all the bonds (x,y) with x ≠ x_0. When we do this one of two things can happen: (a) some path in $\mathcal{B}$ that does not include a bond of the from (x_0,y) is open or (b) there is a set A of neighbors of x_0 (possibly ϕ) so that $\mathcal{B}$ occurs if and only if one of the bonds (x_0,y) is open. The conditional probabilites of $\mathcal{B}$ in the two models are 1 in case (a) and $1-\varphi_1(x_0,A) \le 1-\varphi_2(x_0,A)$ in case (b). Since the joint distribution of the state of the bonds (x,y) with x ≠ x_0 is the same in the two models the result follows.

The argument in the last paragraph proves the result for finite graphs. To pass to the limit to get the result for $\mathbb{Z}^2$, we observe that if $\mathcal{B}$ is a finite set of paths (of finite length) then we can restrict our attention to the vertices that appear in some path, and the result follows from the finite case. If $\mathcal{B}$ is an infinite set of paths (of finite length), then there is a sequence of finite sets $\mathcal{B}_n$ of such paths so that $\mathcal{B}_n \uparrow \mathcal{B}$ as n ↑ ∞. The result for finite sets of paths implies $P_1(\mathcal{B}_n) \le P_2(\mathcal{B}_n)$, and letting n ↑ ∞ gives $P_1(\mathcal{B}) \le P_2(\mathcal{B})$.

If we consider the density of open bonds

$$p = P(\ (0,e_1) \text{ is open }) = 1 - \int_0^\infty e^{-\alpha s/4}\, dF(s)$$

as a parameter, then (2) allows us to identify two extreme cases:

$\varphi_{bond}(A) = (1-p)^{|A|}$ (all bonds independent),

$\varphi_{site}(A) = (1-p)$ if $|A| \ge 1$ (perfect correlation).

Here $|A|$ = the number of points in A. It is clear that any epidemic model with density p has

$$\varphi(A) \le (1-p) = \varphi_{site}(A).$$

To make the other comparison we observe

$$1_{\{ (x,y) \text{ is open} \}} = f(T_x, -e(x,y))$$

where $f(s,t) = 1$ $s+t \ge 0$ and $= 0$ if $s+t < 0$; that is, the indicator functions are increasing functions of independent random variables, so Harris' inequality implies that they are positively correlated. From this, it follows that any epidemic model with density p has

$$\varphi(A) \ge (1-p)^{|A|} = \varphi_{bond}(A).$$

Combining the last two observations with (2) shows that in general

$$P_{site}(\mathscr{B}) \le P(\mathscr{B}) \le P_{bond}(\mathscr{B}).$$

To explain the notation we have used for the extreme cases, we begin by observing that in the "site" epidemic model, either all the bonds (x,y) with $|x-y| = 1$ are open, or all are closed. A little thought reveals that the resulting percolation model is equivalent to the site percolation model discussed in Section 6b. Having heard the term "site" explained the reader can probably guess what the "bond" model refers to, but this time things are not so simple. In the "bond" epidemic model each bond (x,y) with $|x-y| = 1$ is independently open or closed, that is, the states of (x,y) and (y,x) are independent; whereas in the usual model, a bond is either open for passage in both directions or closed. This distinction, however, turns out to be a very minor difference. Frisch and Hammersley (1963) proved a result that implies that the two processes have the same critical value.

(3) Equivalence Theorem. For any two sets of sites S and T the probability there is an open path from S to T is the same in the two models.

Proof: McDiarmid, see (1980) and (1981), obtains this as a special case of the "clutter percolation theorem." We will give a direct proof which does not rely on that general result. The strategy of proof is the same as for (2). We prove the result for finite graphs and then we take limits to get the result for $\mathbb{Z}^2$.

To prove the result for finite graphs, let G be a graph with undirected edges and let G' be the corresponding graph in which all the edges have been replaced by two oriented edges. To interpolate between G and G', number the edges in G, 1,2, ... N, and let G_j be the graph in which all the edges with numbers $\leq j$ have been replaced by two oriented edges (so $G_0 = G$, $G_N = G'$).

To compare G_j and G_{j-1}, observe that if the jth edge is (a,b) then when we split (a,b) into two edges, one of three things can happen: (i) there is a connection from S to T that does not use (a,b) (going in either direction); (ii) there is no path from S to T outside of (a,b) but there are paths from S to a and from b to T that do not use (a,b), so if the oriented edge (a,b) is open there is a path from S to T; (iii) the situation in (ii) occurs and, in addition, there are paths from S to b and from a to T that do not use (a,b), so if (a,b) is open in either direction there is a path from S to T. At first, it looks like case (iii) torpedoes the proof, since the oriented model has two chances to make the connection while the unoriented model has only one. A closer look however reveals that case (iii) is a special case of (i). There are paths from S to b and from b to T that do not use (a,b), so there is a path from S to T that does not use (a,b). With (ii) out of the way, we are left with case (ii) in which each model has exactly one chance to make the connection and each will succeed with probability p.

Combining (2) and (3) with the observations in between, it follows that if we let

$$\alpha_c(F) = \inf \{ \, \alpha : P_{\alpha,F}(\, C_0 \text{ is infinite }) > 0 \, \},$$

and

$$p_c(F) = 1 - \int_0^\infty \exp(-s \, \alpha_c(F)/4) \, ds,$$

then

$$1/2 \leq p_c(F) \leq p_c(\text{site}) \approx .5927,$$

the numbers at the left and right being the rigorous and numerical critical values for bond and site percolation respectively.

The results in the last paragraph (which are due to Kuulasmaa (1982)) show that $0 < \alpha_c(F) < \infty$, and if you believe the .5927, give reasonable bonds on the critical value. Having established the existence of a phase transition, we turn next to the question: What does the epidemic look like when it lasts forever? Before answering this question, we start with a simpler problem.

Consider the usual bond percolation model in which bonds are open (in both directions) with probability p, closed with probability 1–p, and distinct bonds are independent. If we think of open bonds as being matchsticks that take exactly one unit of time to burn, and we set the origin on fire at time 0, then at time n all the sites in C_0 at a distance less than n from 0 will be burnt, and those at distance n will be on fire. If we let B_n = the set of sites burnt at time n, and F_n = the set of sites on fire at time n, then the shape theorem for this model may be stated as:

(4) Theorem. If p > 1/2 then there is a convex set A such that for any $\epsilon > 0$, as n → ∞

$$P(\ C_0 \cap n(1-\epsilon)A \subset B_n \subset n(1+\epsilon)A\ ) \to 1,$$

and

$$P(\ F_n \subset n(1+\epsilon)A - n(1-\epsilon)A\ ) \to 1.$$

In words, if n is large, then with high probability, all the sites in $n(1-\epsilon)A$ that will ever burn are already burnt, and the fire is near $n(\partial A)$.

Given results in Section 8a, this is not hard to prove. Consider first passage percolation in which undirected bonds [x,y] are independently assigned passage time 1 with probability p, and ∞ with probability 1−p. (Notice the square brackets. Here and in what follows, they signal that we are dealing with unoriented bonds.) It is easy to see that if we let t[x,y] be the passage time from x to y, then $B_n = \{\ x : t[0,x] < n\ \}$, and $F_n = \{\ x : t[0,x] = n\ \}$, so the burning percolation cluster fits into the first passage percolation set up.

The results of Section 8a cannot be directly applied to the model above, because in that section we assumed t[x,y] < ∞. This difficulty is easy to remedy, because at the beginning of the proof of part (b) of (1), we picked M so that F(M) = p > 1/2, called bonds with passage time > M closed, and then bounded passage times by constructing paths that consist only of open bonds. Given this general strategy, the reader should not find it too hard to believe that the proofs given in Section 8a generalize immediately to prove (4). We will not give any details because Y. Zhang and Y.C. Zhang (1984) have already proved a closely related result — they considered first passage percolation with P(X[x,y] = 0) = p > 1/2, P(X[x,y] = 1) = 1−p, and studied the asymptotic behavior of the length of the shortest path from 0 to x that achieves the passage time t[0,x].

Returning to the forest fire, let $B_t = \{ x : \xi_t^0(x) = 0 \}$ and $F_t = \{ x : \xi_t^0(x) = i \}$ be the set of trees that are burnt and on fire (respectively) at time t, when initially we start with one burning tree at 0 sitting in the middle of an otherwise virgin forest. The "shape theorem" is:

(5) Theorem. Assume that $\int_0^\infty x^2 \, dF(x) < \infty$, and $\alpha > \alpha_c(F)$. Then there is a convex set A so that for any $\epsilon > 0$, as $t \to \infty$

$$P(\, C_0 \cap t(1-\epsilon)A \subset B_t \subset t(1+\epsilon)A \,) \to 1,$$

and

$$P(\, F_t \subset t(1+\epsilon)A - t(1-\epsilon)A \,) \to 1.$$

The second moment assumption is necessary for the conclusions above. Otherwise, repeating the part of the proof of (8) from Section 8a, shows that for any $K < \infty$, we will infinitely often encounter trees with $|x| < t$ which burn for more than Kt units of time.

The proof of (5), like that of (4), is a question of analyzing first passage percolation in which some of the bonds have infinite passage time. The hardest part of doing this is to develop a theory of percolation in locally dependent random graphs which parallels that developed in Section 6a. Once this is done, it is reasonably straightforward to follow the approach of Section 8a to prove (5), so we just give the details of the hard part and leave the reader to work out the rest, or look up the answer in Cox and Durrett (1988).

The first step in developing the percolation theory necessary to study the forest fire is to generalize (2) of Section 6a. Let $\rho_{m,n}$ be the probability there is a right to left crossing of $(0,m) \times (0,n)$ by open bonds. Notice that orientation is now important.

$$(6) \qquad\qquad \rho_{3L/2,L} \geq (1 - (1 - \rho_{L,L})^{1/2})^3.$$

Proof: The proof starts like it did before. Refer to Figure 9.1 for help with the definitions. Let E_s be the event that s is the lowest right to left crossing of $(0,L) \times (0,L)$. Let s_r be the portion of the path from $\{L\} \times (0,L)$ until the first time it hits $\{L/2\} \times (0,L)$. Let s_{rr} be the reflection of s_r through $\{L\} \times (0,L)$. Let $\mathscr{A}(s_r \cup s_{rr})$ be the set of points in $(L/2,3L/2) \times (0,L)$ above $s_r \cup s_{rr}$, and let $\mathscr{B}(s) =$ the points in $(0,L) \times (0,L)$ on, or below s.

For reasons that we will explain in a minute, the next step must be different. For all $x \in \mathscr{B}(s)$ (and y with $|x-y| = 1$), define new variables T'_x, $e'(x,y)$, $X'(x,y)$ which are independent of the original random variables and have the same distribution, and extend the definition of the primed variables to $(0,3L/2) \times (0,L)$ by setting $T'_x = T_x$, $e'(x,y) = e(x,y)$, $X'(x,y) = X(x,y)$ for $x \notin \mathscr{B}(s)$. For convenience of exposition, we will refer to the variables with primes as new variables, and call the original variables old.

Since E_s is measurable with respect to the sigma field generated by the variables on bonds that begin in $\mathscr{B}(s)$, the new variables have the same distribution as the old ones. Let F'_s be the event that there is a path t in the new system starting from $(L/2, 3L/2) \times \{L\}$ and connected to s_r in $\mathscr{A}(s_r \cup s_{rr})$. In Russo's original proof, this path could be combined with s to make a path to s_r in the old system but this time the orientation is wrong. (See Figure 9.1.) Fortunately there is a way around this.

Let t' consist of the bonds in t that begin in $\mathscr{A}(s_r \cup s_{rr}) - \mathscr{B}(s)$. We claim

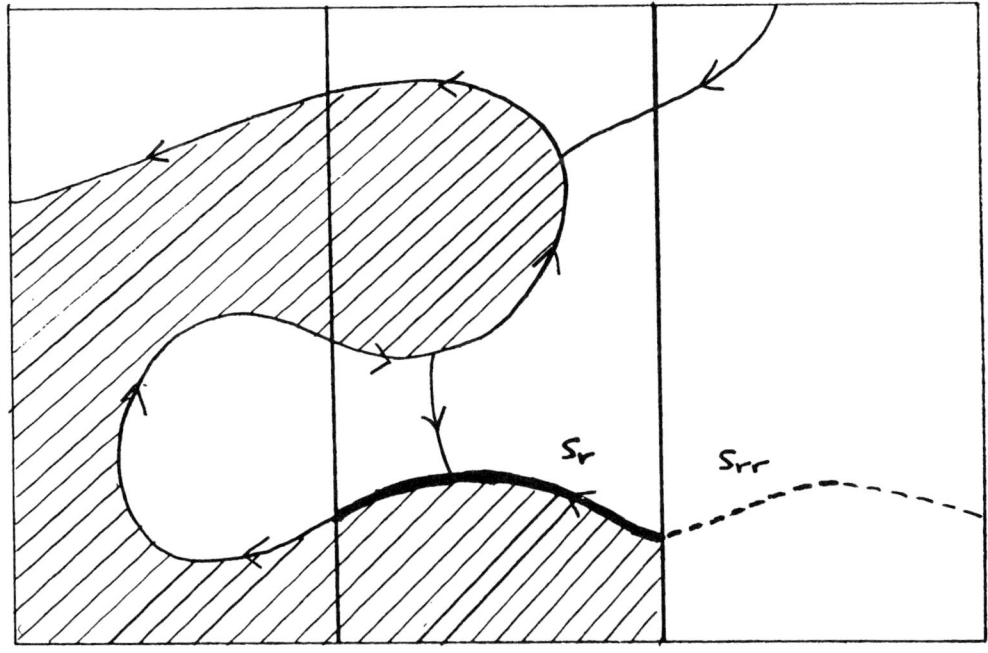

Figure 9.1

that the union of s and t' "protects {L/2} × [L/2,L]." That is, if u is a right to left

crossing of (L/2, 3L/2) × (0,L) which ends in {L/2} × [L/2, L], then s, t', and u can be

combined to give a right to left crossing of (0,3L/2) × (0,L). We have not drawn a path

like u in Figure 9.1, but invite you to do so now. The Jordan curve theorem implies

that u must intersect s or t. If u intersects s, then following u until it first hits s and

then continuing along s gives the desired path. If u intersects t and the first

intersection (along u) is in $\mathscr{B}(s)$, then we are done, since u must have intersected s at

an earlier ($\leq$) time. If, on the other hand, the first intersection is in $\mathscr{B}(s)^c$, we follow

t' from this point until the next time it hits s and continue with s to cross the

rectangle.

     Having shown that u, t', and s can be combined to give the desired path, most of the rest of the proof is the same as before. At the end of this section, we will need to generalize this argument one more time, so before we finish the argument the reader should pause a moment to notice that up to this point the proof has been concerned mostly with the "geometry" of oriented paths, and has used very little probability.

Returning to our original program, let G be the union of $E_s \cap F'_s$ over all paths for which the last point of s_r has y coordinate $\leq L/2$. Let H be the event that there is right to left crossing of $(L/2, 3L/2) \times (0,L)$ which ends at a point with y coordinate $\geq L/2$. Since the occurrence of G and H guarantees the desired crossing, it suffices to show that

$$P(G \cap H) \geq (1 - (1 - \rho_{L,L})^{1/2})^3.$$

Using Harris' inequality and the square root trick, gives as before that

$$P(G \cap H) \geq P(G)\, P(H),$$

and

$$P(H) \geq (1 - (1 - \rho_{L,L})^{1/2})^3.$$

To estimate P(G), we write

$$P(G) = \sum_s P(E_s \cap F'_s) = \sum_s P(E_s)P(F'_s | E_s),$$

and use the square root trick with $A_1 = F'_s$ and $A = \{$ there is a path from $(L/2, 3L/2) \times \{L\}$ down to $s_r \cup s_{rr}$ in $\mathscr{A}(s_r \cup s_{rr})$ in the "new" system $\}$ to conclude

$$P(F'_s | E_s) \geq 1 - (1 - P(A | E_s))^{1/2} \geq 1 - (1 - \rho_{L,L})^{1/2}.$$

Here, we are using the fact that if we condition on E_s and use the new variables, then the $X'(x,y)$ with $x \in \mathcal{A}(s_r \cup s_{rr})$ are independent, and have the same distribution as the old variables. This is the step for which the primed variables were introduced. As we remarked in Section 6a, this is one of two places where independence is used. With the last inequality in hand the proof is complete, because it implies

$$P(G) \geq (1 - (1 - \rho_{L,L})^{1/2}) \sum_{s} P(E_s),$$

and another use of the square root trick gives the desired result.

With (6) established, the rest of the developments are almost exactly as in Section 6a. We draw a couple of pictures (see Figures 9.2 and 9.3), and use the trivial

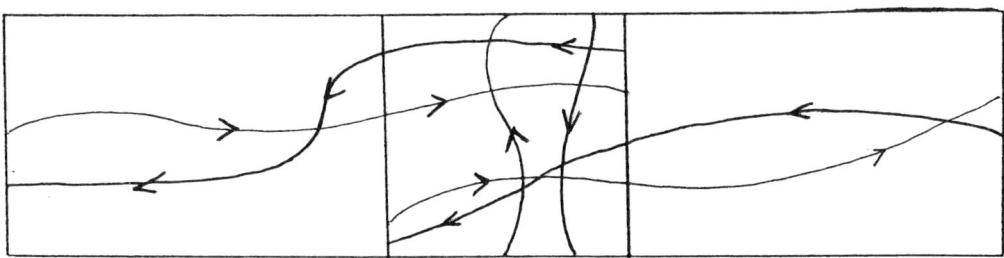

Figure 9.2

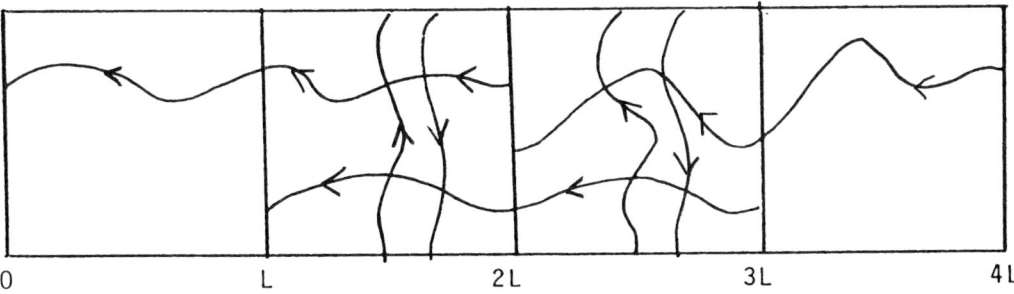

0   L   2 L   3 L   4 L

Figure 9.3

inequality

$$P(\bigcap_{i=1}^{k} A_i) \geq 1 - \sum_{i=1}^{k} P(A_i^c)$$

to conclude:

(7) $$\rho_{kL,L} \geq 1 - 4(1 - \rho_{(k+1)L/2,L}).$$

(8) $$\rho_{4L,L} \geq 1 - 7(1 - \rho_{2L,L}).$$

Then we put two 4L by L rectangles next to each other to conclude

(9) $$\rho_{4L,2L} \geq 1 - (1 - \rho_{4L,L})^2.$$

As we remarked in Section 6a, this is the second place where independence is used, but (9) is valid here since the existence of crossings in $(0,4L) \times (0,2L)$ and $(0,4L) \times (L,2L)$ are independent.

Combining (8) and (9) gives

(10) $$\rho_{4L,2L} \geq 1 - 49(1 - \rho_{2L,L})^2.$$

If $\rho_{2L,L} = 1 - \lambda/49$ with $\lambda < 1$, then iterating turns the last result into

(11) $$\rho(2^k L, 2^{k-1} L) \geq 1 - (1/49) \exp(2^{k-1} \log \lambda).$$

Combining the last result with (6) and (7) gives

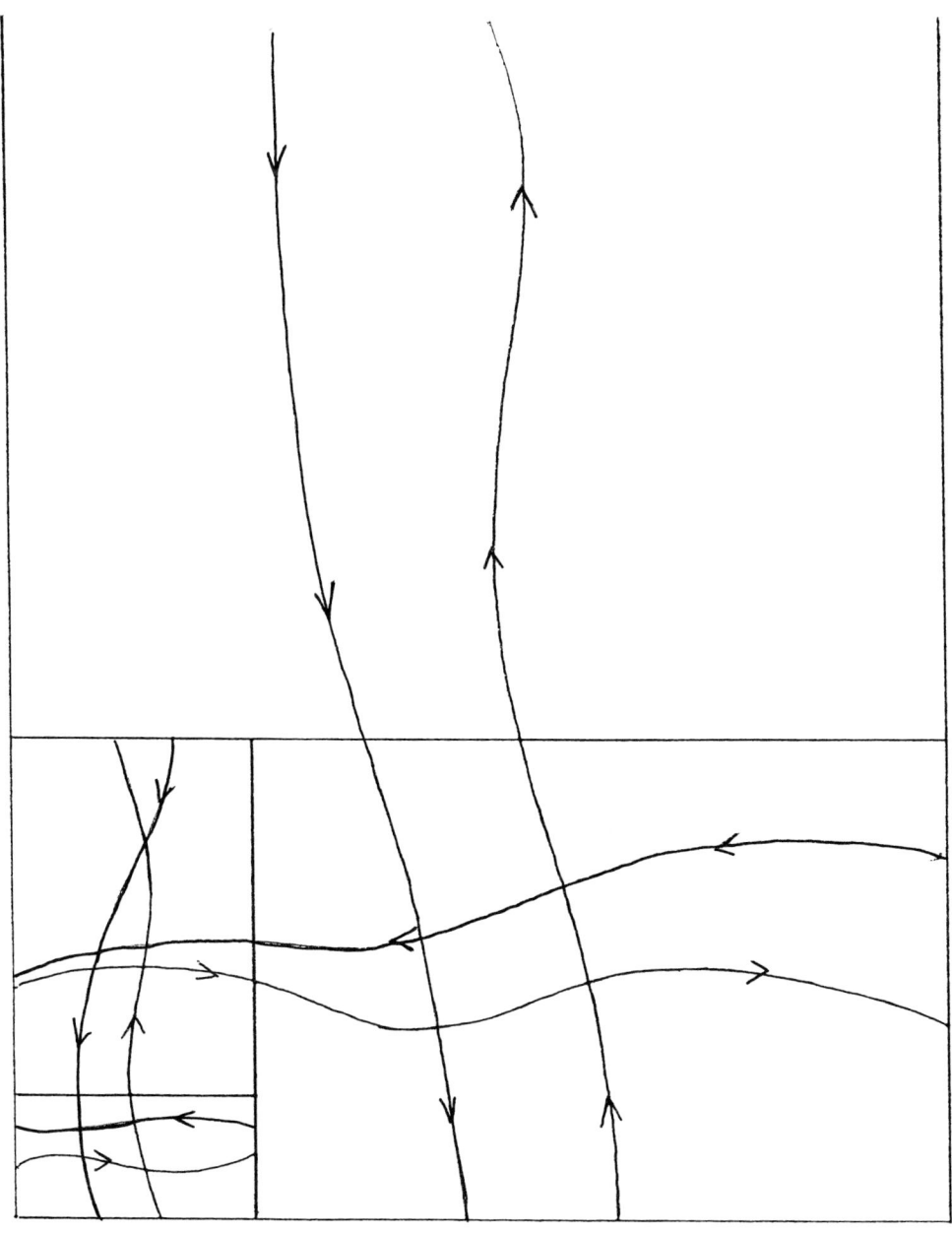

Figure 9.4

(12) Rescaling lemma. There is an ϵ_0 so that if $\rho_{L,L} > 1 - \epsilon_0$, then

$$\rho(2^k L, 2^{k-1} L) \geq 1 - (1/49) \exp(-2^{k-1}).$$

(Pick ϵ_0 so that $\rho_{2L,L} \geq 1 - 1/49e$.) With (12) established, we can now produce percolation much as we did in Section 6a. Define boxes

$$B_{2k-1} = (0, 2^{2k-1} L) \times (0, 2^{2k-2} L),$$
$$B_{2k} = (0, 2^{2k-1} L) \times (0, 2^{2k} L),$$

and observe that if we get left to right and right to left crossings of all the B_{2k-1}, and top to bottom and bottom to top crossings of all the B_{2k} (no you are not not seeing double double), then there is an infinite path starting on $\{0\} \times (0,L)$. For a picture, see Figure 9.4. The probability we get all the paths we want, and in addition that all the bonds on $\{0\} \times (0,L)$ are open is, by Harris' inequality, at least

$$p^{2L} \prod_{k=1}^{\infty} [1 - (1/49)\exp(-2^{k-1})]^2 > 0.$$

The developments above should motivate defining

$$\alpha_s = \inf \{ \alpha : \sup_L \rho_{L,L} > 1 - \epsilon_0 \},$$

where s stands for "sponge crossing" and ϵ_0 is the magic constant in (12). The critical value just defined could be different from α_c defined above. The next result shows that this is not the case, and that the forest fire dies out at the critical value.

(13) Theorem. $\alpha_s = \alpha_c$ and $P_{\alpha(s)}(\, |C_0| = \infty \,) = 0$.

Proof: We have already shown that $\alpha_c \le \alpha_s$. To prove the other inequality, observe that if $\alpha < \alpha_s$ then $\rho_{L,L}(\alpha) \le 1 - \epsilon_0$ for all L. By continuity, the last conclusion implies $\rho_{L,L}(\alpha_s) \le 1 - \epsilon_0$ for all L. With the probabilities of sponge crossings bounded away from 1, we can use the argument of Harris (1960) that we used in the proof of (9) in Section 6b to show that there is no percolation. To carry out the proof we need to generalize a few things from Chapter 6.

Introduce the dual percolation process with sites $Y^2 = (1/2,1/2) + \mathbb{Z}^2$, and call the bond (u,v) between neighboring points in Y^2 open (closed) if the bond on the original lattice obtained by rotating it 90 degrees (counterclockwise) around its midpoint is open (closed). Generalizing the duality used in the unoriented case gives:

(14) Lemma. Either there is a right to left crossing of $(0,L) \times (0,L)$ by open bonds, or top to bottom crossing of $[1/2,L-1/2] \times [1/2, L-1/2]$ by closed bonds on the dual, but not both.

Proof: For completeness we give a proof, but the reader can safely skip it. First, if there is a top to bottom crossing on the dual, then there is a self–avoiding one (i.e., one in which each site appears at most once.) If we call the path σ, an application of the Jordan curve theorem shows that σ divides the interior of the square into two parts – one we call T_1, which lies to the right of S, and another we call T_2, which lies to the left of σ. If we move along σ in the direction of the orientation, then T_1 is always on our left, and T_2 is always on our right. From this, we see that if there is a right to left crossing by open bonds, then any time it crosses from T_1 to T_2 it does so along a bond that is a 90 degree clockwise rotation of a bond on σ. But such bonds are closed, so no

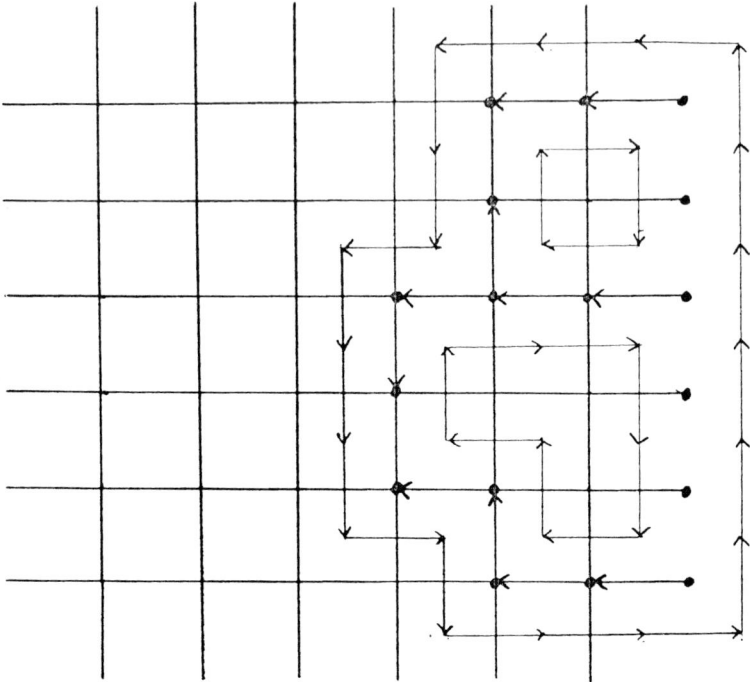

Figure 9.5

right to left open crossing exists.

To prove the other direction, we will suppose there is no right to left crossing and construct a top to bottom one. The reader should consult Figure 9.5 for help with the definitions below. Let C be the set of points that can be reached from the right edge by a path of open bonds. Let $D = \{ (a,b) : |a|,|b| \leq 1/2 \}$, and orient the boundary of D in a counterclockwise fashion. Finally, let $W = \cup_{z \in C} (z + D)$. If we combine the boundaries of the $z + D$ with $z \in C$, and let oppositely directed segments cancel, then the boundaries that remain are closed paths on the dual. One of these paths, Γ = the boundary of the component of $(0,L) \times (0,L)$ that contains the left side of the box is the path we want, and the proof of (14) is complete.

(14) implies that the probability of a top to bottom crossing of $[1/2, L-1/2] \times$ $[1/2, L-1/2]$ by closed bonds is bounded away from 0 when $\alpha = \alpha_s$. The next step is to show that Russo's lemma can be generalized to the dual model. To see that this is legitimate, observe that in the dual the bonds $(m,n) \to (m+1,n) \to (m+1,n+1) \to$ $(m,n+1) \to (m,n)$ are dependent but bonds that go counterclockwise around different squares are independent. From the last observation, we see that if s is the lowest right to left crossing, then all the bonds above s are independent of it and the argument given for (6) works in this case.

With Russo's lemma generalized to the dual model, the rest of the proof of (13) follows the proof of $p_c \geq 1/2$ in Section 6a. Drawing a picture (see Figure 9.2 and turn your head sideways) and using Harris' inequality shows that the probability of a top to bottom crossing of an $L \times kL$ rectangle is bounded away from 0 for any $k < \infty$.

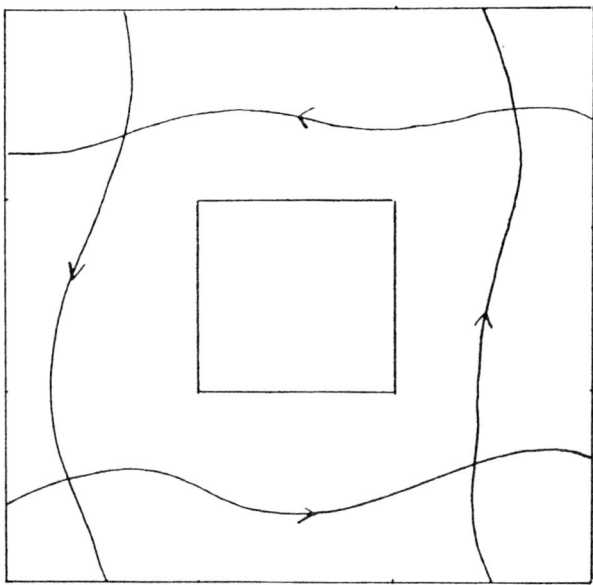

Figure 9.6

Combining two crossings of L × 3L and two crossings of 3L × L rectangles (see Figure 9.6), shows that a closed circit of dual bonds in an annulus has a probability bounded away from 0, and surrounding the origin with infinitely many disjoint annuli shows that the probability of percolation is 0.

With the proof of (13) completed, this section comes abruptly to an end. We have developed the percolation theory necessary to generalize the proof of (8) in Section 8a, and we leave it to the reader to do this exercise or look up the answer. We close this section with some simulations of the process with $F(x) = 1_{(x \geq 1)}$. We have chosen this case because when the disease lasts for exactly one unit of time, the states of the four bonds out of a site are independent, and it follows from (3) that $p_c = 1/2$. There are three sets of three pictures. In each set, the process is shown the first time the fire exits the boxes with radii 20, 40, and 80, and burnt sites are black. In the first set of pictures, p = .6, and as the theorem predicts, the times roughly double from one picture to the next. In the second set, p = .75. Here, as in the discrete time version of Richardson's model, the limiting shape has a "flat edge." The proof is the same as the one in Section 5a. If we look at the set of points on x + y = n that are on fire at time n, then what we see looks like oriented percolation. The last set has p = .51. The last picture does not look much like a ball, but that is to be expected. The distance that the process has to spread before it begins to look like a ball is one way of defining the "correlation length" for this model (see the discussion of (8) in Section 8b), and that length scale diverges as $p \downarrow p_c$. Physicists tell us that if we look at the picture the first time the fire reaches the correlation length and let $p \downarrow p_c$, then the system (scaled in the obvious way) converges in distribution to a random fractal with Hausdorff dimension ≈ 1.9.

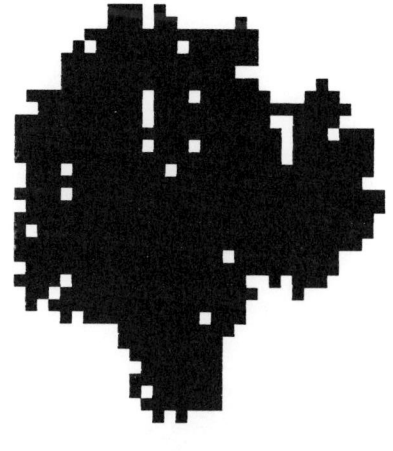

TIME 21

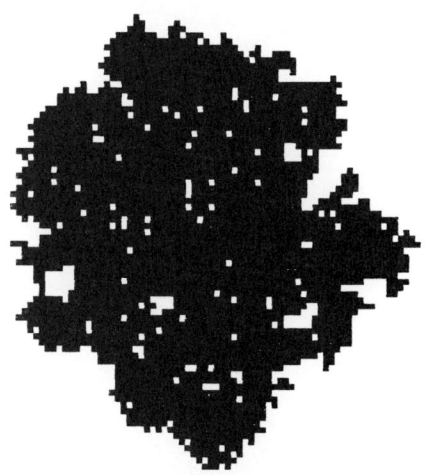

TIME 47

TIME 105

P = .60

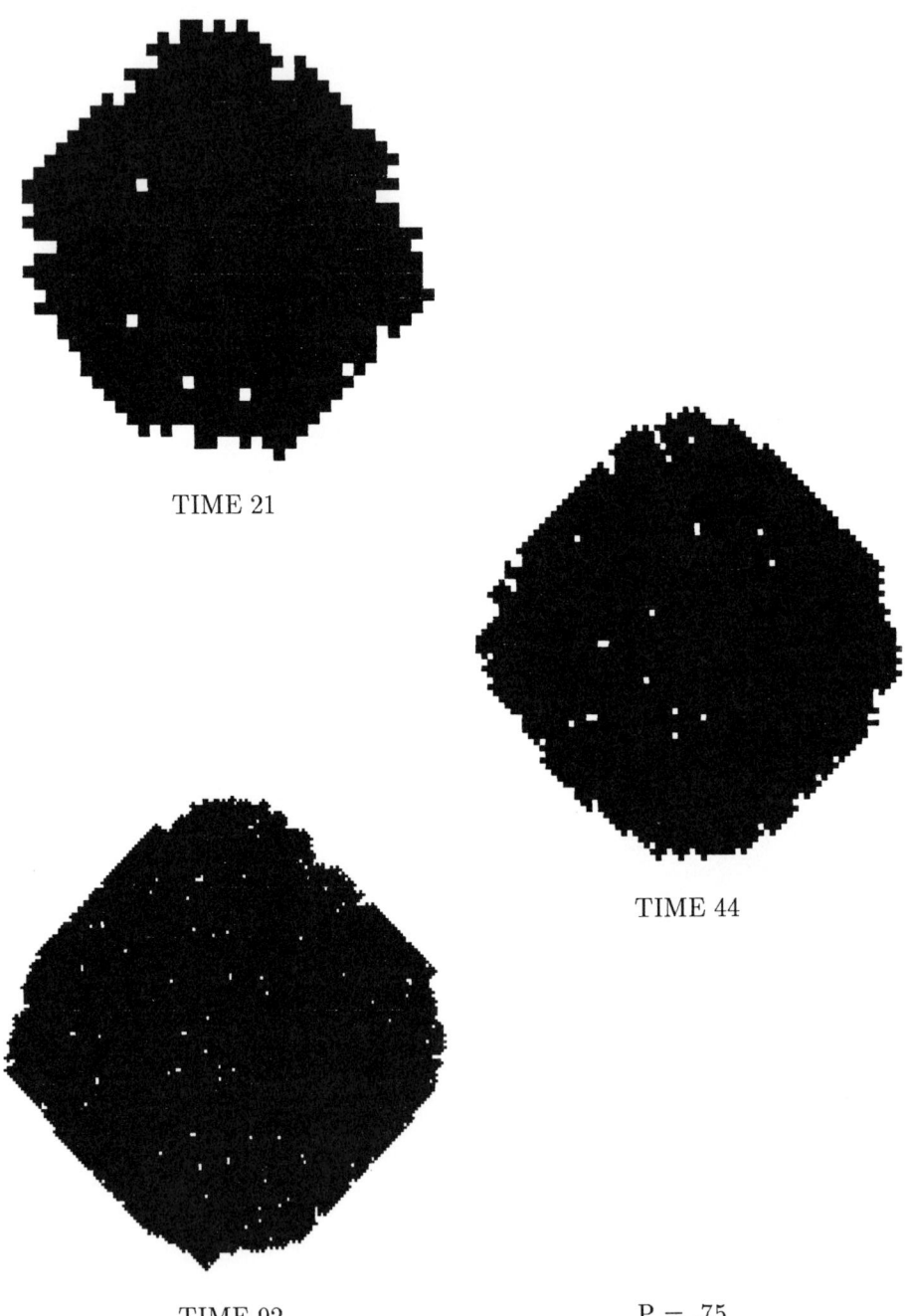

TIME 21

TIME 44

TIME 92   P = .75

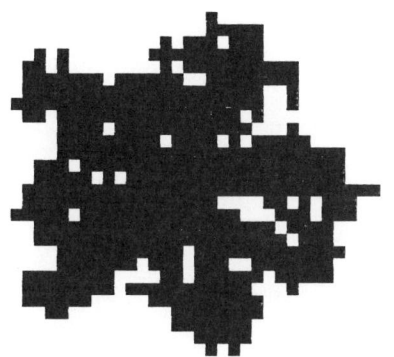

TIME 21

TIME 57

TIME 141   P = .51

10 The Voter Model, II

10a Clustering in d = 1 and d = 2

In Chapter 2, we saw that in $d \le 2$ the voter model approached complete consensus. In this section, we take a look at how the clusters of individuals with the same opinion grow as time goes on. Perhaps the most basic question is: How big are the clusters at time t? We begin with one dimension, where the clustering is easy to conceptualize and analyze mathematically, and then proceed to two dimensions where the answers are as beautiful as they are difficult to prove.

In this section, we only consider the voter model starting from product measure with density θ, and use ξ_t^θ to denote the process starting from that initial configuration. In the next result, and many times below, we should replace $x\sqrt{t}$ by $[x\sqrt{t}]$, that is, the integer part of $x\sqrt{t}$, but having mentioned our "error," we will leave it to the reader to correct it.

(1) **Theorem.** Let $d = 1$. As $t \to \infty$

(i)
$$P(\xi_t^\theta(x\sqrt{t}) \ne \xi_t^\theta(y\sqrt{t})) \to 2\theta(1-\theta) \left[2 \int_0^{|x-y|} \frac{e^{-z^2/4}}{\sqrt{4\pi}} \, dz \right],$$

and (ii) the finite dimensional distributions of $\{ \xi_t^\theta(x\sqrt{t}) : x \in \mathbb{R} \}$ converge to those of a limit field $\{ \xi_\infty^\theta(x) : x \in \mathbb{R} \}$.

Proof: The first statement is easy to prove. By duality

$$P(\xi_t^\theta(x\sqrt{t}) \neq \xi_t^\theta(y\sqrt{t})) = 2\theta(1-\theta)P(\zeta_t^{x\sqrt{t}} \neq \zeta_t^{y\sqrt{t}}).$$

Before the particles hit, the difference $\zeta_s^a - \zeta_s^b$, $s \geq 0$ is a simple random walk starting from $a-b$ and run at rate 2. (Sticklers for details will object that ζ_s^a and ζ_s^b are sets, but here and in what follows, we will ignore the difference between $\{x\}$ and x.) From the last observation and Donsker's theorem, it follows that

$$P(\zeta_t^{x\sqrt{t}} \neq \zeta_t^{y\sqrt{t}}) \to P_{x-y}(B_s \neq 0 \text{ for all } s \leq 2),$$

where B_s is a Brownian motion. If $x < y$, the reflection principle tells us the last probability is $P(B_2 \geq x-y) - P(B_2 \geq y-x) = P(|B_2| \leq y-x)$.

To prove (ii), we observe that, as in Chapter 2, it suffices to prove that $P(\xi_t^\theta \cap B = \phi)$ has a limit for all $B \subset \{x_1\sqrt{t},\ldots,x_n\sqrt{t}\}$. Since

$$P(\xi_t^\theta \cap B = \phi) = E((1-\theta)^{|\zeta_t^B|}),$$

it is enough to show that for any $y_1 < y_2 \ldots < y_m$ and $1 \leq k \leq m$

$$P(|\zeta_t(y_1\sqrt{t},\ldots,y_m\sqrt{t})| = k)$$

has a limit, where $\zeta_s(y_1\sqrt{t},\ldots,y_m\sqrt{t})$ is the dual process starting from $\{y_1\sqrt{t},\ldots,y_m\sqrt{t}\}$ occupied. To get started on this, we observe that another application of Donsker's theorem implies that as $t \to \infty$, $P(|\zeta_s(y_1\sqrt{t},\ldots,y_m\sqrt{t})| = m)$ converges to the probability a Brownian motion starting from $(y_1,\ldots,y_m)$ in $\mathbb{R}^m$ has not left $\{z : z_1 < \ldots < z_m\}$ by time 1. To deal with $P(|\zeta_t(y_1\sqrt{t},\ldots,y_m\sqrt{t})| = k)$ for $k < m$, let $\tau = \inf\{ s :$

Figure 10.1

$|\hat{\xi}_s(y_1\sqrt{t},...,y_m\sqrt{t})| < m\}$, break things down according to the value of τ and the positions of the particles at that time, and use induction.

The last result proves the existence of a limit but does not say much about it. In describing ξ_∞^θ, a picture is worth at least a thousand words (see Figure 10.1). To describe the picture, it is useful to introduce an auxillary system which is of interest in its own right. Define $\eta_t \subset \mathbb{Z}+1/2$ by putting $x \in \eta_t$ if and only if $\xi_t(x-1/2) \neq \xi_t(x+1/2)$ (where as before, $\xi_t(y) = 1$ if $y \in \xi_t$, and 0 if $y \notin \xi_t$). The points in η_t mark the boundaries between different opinions in the voter model. To see how η_t evolves in time, consider two examples of what can happen when $x+1/2$ is in η_t and the voter at x imitates the voter at $x+1$. (See Figure 10.2.). As usual, we will call the points in η_t particles. In the first picture, the particle at $x+1/2$ jumps to $x-1/2$. In the second picture, there is a particle at $x-1/2$ and the two particles disappear; or as we prefer to say, the particle at $x+1/2$ jumps to $x-1/2$, and when it lands on the other particle, the two particles annihilate each other.

	x	x
before	0.1 1 1.0 0	0 0.1.0 0
after	0.1 1.0 0 0	0 0 0 0 0

Figure 10.2

With η_t introduced, we can describe the picture in a few words; "the boundaries between different opinions perform independent random walks." From this perspective, it is also easy to explain the limit theorem in (ii). As $t \to \infty$, $\eta_{st}^\theta/\sqrt{t}$, $s \geq 0$ converges to a set of annihilating Brownian motions. This is not too hard to visualize when s > 0. It is better not to try to think about what happens at s = 0. In the limit, the points of $\eta_0^\theta/\sqrt{t}$ become dense in $\mathbb{R}$, but a "big bang" occurs at s = 0, and the number of points in any bounded set is finite when s > 0.

(2) Theorem. Let d = 2. If $0 < \alpha < 1$ then as t → ∞

(i)
$$P(\xi_t^\theta(xt^{\alpha/2}) \neq \xi_t^\theta(yt^{\alpha/2})) \to 2\theta(1-\theta)\alpha,$$

and (ii) the finite dimensional distributions of $\{ \xi_t^\theta(xt^{\alpha/2}) : x \in \mathbb{R}^2 \}$ converge to those of a limit field $\{ \xi_\infty^{\theta,\alpha}(x) : x \in \mathbb{R}^2 \}$.

The reader's first reaction to this result should be that it cannot possibly be true! In (1), as in most limit theorems, there is only one way to scale the process to get an interesting limit. There, if we scale space by f(t) with $f(t)/t^{1/2} \to 0$, then the limit is perfectly correlated, and if $f(t)/t^{1/2} \to \infty$, the sites are independent in the limit.

Proof of (2): Let me start by admitting that I'm not actually going to prove (2), but just indicate the ideas that go into the proof. I hope you will *go with the flow*, and not worry too much about how one line follows from the next. More details than you want to read can be found in Cox and Griffeath (1986).

To "prove" (i), we observe that the difference process

$$X_s = \tilde{\xi}_s(x_1 t^{\alpha/2}) - \tilde{\xi}_s(x_2 t^{\alpha/2}), \quad s \geq 0$$

is a rate 2 simple random walk until the particles meet, so

(3) $$P(|\xi_t(x_1 t^{\alpha/2}, x_2 t^{\alpha/2})| = 1) = P_{xt^{\alpha/2}}(\tau_0 \leq t),$$

where $x = x_1 - x_2$, $\tau_0 = \inf\{ s \geq 0 : X_s = 0 \}$, and the subscript indicates the starting point. Breaking things down according to the value of $\sigma_t = \sup\{ s \leq t : X_s = 0 \}$, and writing $p_s(x,y) = P_x(X_s = y)$ gives

(4) $$P_{xt^{\alpha/2}}(\tau_0 \leq t) = p_t(xt^{\alpha/2},0) + \int_0^t p_u(xt^{\alpha/2},0) \, P_{(1,0)}(\tau_0 > t-u) \, du.$$

Now X_s is a simple random walk moving at rate 2, and each component of each step has variance $1/2$, so the local central limit theorem implies

(5) $$p_u(xt^{\alpha/2},0) \approx \frac{1}{2\pi u} e^{-|x|^2 t^{\alpha}/2u},$$

if u is large. Consulting Spitzer (1976), we see

(6) $$P_{(1,0)}(\tau_0 > t-u) \approx 2\pi/\log(t-u)$$

if t−u is large.

   When the smoke clears, the expression in (4) will converge to a positive limit so we can throw away the first term on the right, and the integral from 0 to 1. Plugging (5) and (6) into what remains gives

$$\int_1^t \frac{1}{u} e^{-|x|^2 t^{\alpha}/2u} \, \frac{1}{\log(t-u)} \, du.$$

(5) cannot be believed when u is near 1, and (6) is wrong when u is near t. Neither of these problems is serious, and the reader can easily fix them when she sees where we are going. Continuing our computation, we change variables $\gamma = (\log u)/\log t$ (i.e., u $= t^{\gamma}$) to get

$$\int_0^1 e^{-|x|^2 t^{\alpha-\gamma}/2} \frac{\log t}{\log(t-t^{\gamma})} \, d\gamma.$$

As $t \to \infty$, the first factor $\to 0$ for $\gamma < \alpha$, and $\to 1$ for $\gamma > \alpha$, while the second $\to 1$ for all $\gamma < 1$. After truncating to take care of γ near 1 (recall the quantity estimated in (6) is a probability), we conclude.

$$P(|\tilde{\xi}_t(x_1 t^{\alpha/2}, x_2 t^{\alpha/2})| = 1) \to \int_\alpha^1 1 \, d\gamma = 1-\alpha.$$

This proves part (i) of (2) (modulo a few details). The reader should note that the limit depends on α but is independent of x_1-x_2.

To prove (ii) it suffices, as in the proof of (1), to show

(7) $\lim_{t \to \infty} P(|\tilde{\xi}_t(x_1 t^{\alpha/2},...,x_n t^{\alpha/2})| = k)$ exists.

We call the limit $p_{n,k}(\alpha)$ and prove the result by induction. The argument above handles the case n = 2. The next step is to show for n $\geq$ 3 and $0 < \alpha \leq \beta < \infty$.

(8) $P(|\hat{\xi}_t^\beta(x_1 t^{\alpha/2},...,x_n t^{\alpha/2})| = n) \to (\alpha/\beta)^{\binom{n}{2}}.$

We will not indicate the proof here, but merely point out that it is the right answer if one believes that the events

$$\{\,|\hat\xi_{t^\beta}(x_i t^{\alpha/2}, x_j t^{\alpha/2})| = 2\,\}\quad\text{are independent,}$$

since by the first result, each of these $\binom{n}{2}$ events has probability α/β. It would be nice to have a proof that makes this intuition precise. The proof of Cox and Griffeath (1986) (see p.355 and Section 5) is, as they admit, "rather devious."

For the next step, it is useful to observe that if τ is the time of the first collision, then (8) says

$$P(\tau \geq t^\beta) = (\alpha/\beta)^{\binom{n}{2}}\quad\text{for } \beta \geq \alpha.$$

To handle $k < n$, we observe

$$P(|\hat\xi_t(x_1 t^{\alpha/2},...,x_n t^{\alpha/2})| = k)$$
$$= \int_0^t P(\tau \in ds,\text{ the remaining n}-1\text{ particles coalesce to k particles by time t}).$$

If we change variables $s = t^\gamma$, assume that the remaining $n-1$ particles at time t^γ are spaced about $t^{\gamma/2}$ apart, and recall that their exact spatial location is unimportant, then the last result and the induction hypothesis imply that

(9)
$$p_{n,k}(\alpha) = \int_\alpha^1 \binom{n}{2}\, \alpha^{\binom{n}{2}}\, \gamma^{-\binom{n}{2}-1}\, p_{n-1,k}(\gamma)\, d\gamma.$$

The last expression relates $p_{n,k}(\alpha)$ to $p_{n-1,k}(\alpha)$. $p_{k,k}(\alpha)$ is known, so the $p_{n,k}(\alpha)$ can be computed by induction. The answer is

$$
(10) \qquad p_{n,k}(\alpha) = \sum_{j=k}^{n} (-1)^{j+k} \frac{(2j{-}1)(j{+}k{-}2)!}{k!(k{-}1)!\,(j{-}k)!} \frac{\begin{bmatrix} n \\ j \end{bmatrix}}{\begin{bmatrix} n{+}j{-}1 \\ j \end{bmatrix}} \alpha^{\binom{j}{2}},
$$

so we will spare the reader the details of the verification. As we pointed out earlier, the convergence of the probabilities in (10) is sufficient to prove (ii), so we have completed the proof of (2).

While formula (10) is stunning, it is probably not what you would call pretty (unless you find !'s exciting). The beauty in the solution begins to emerge when we find a simpler description for the $p_{n,k}(\alpha)$. To do this, we differentiate the integral formula (9) to get

$$
\frac{d}{d\alpha} p_{n,k}(\alpha) = \binom{n}{2} \int_{\alpha}^{1} \binom{n}{2} \alpha^{\binom{n}{2}-1} \gamma^{-\binom{n}{2}-1} p_{n-1,k}(\gamma) d\gamma
$$
$$
- \binom{n}{2} \alpha^{-1} p_{n-1,k}(\alpha),
$$

and change variables $\alpha = e^{-u}$ to get

$$
(11) \qquad \frac{d}{du} p_{n,k}(e^{-u}) = -\binom{n}{2} p_{n,k}(e^{-u}) + \binom{n}{2} p_{n-1,k}(e^{-u}).
$$

(Recall $d/du = (d\alpha/du)(d/d\alpha) = (-\alpha)(d/d\alpha)$.) If we let $q_{n,k}(u) = p_{n,k}(e^{-u})$, then we see that $q_{n,k}(u) = P_n(\,D_u = k\,)$ where D_u is a "pure death process" on the positive integers which jumps from j to $j{-}1$ to rate $\binom{j}{2}$ for $j \geq 2$. So

(12) $$p_{n,k}(\alpha) = P_n(D_{\log(1/\alpha)} = k).$$

The last formula gives an expression for the $p_{n,k}(\alpha)$ that is not as explicit as (10), but certainly easier to remember. Our next goal is to describe the limit field $\{\xi_\infty^{\theta,\alpha}(x) : x \in \mathbb{R}^2\}$. The absence of spatial dependence in the limiting correlations means that the limit field is exchangeable, and hence, de Finetti's theorem asserts that this field is a mixture $dF_{\theta,\alpha}(s)$ of Bernoulli product measures with density s. In particular,

(13) $$P(\xi_\infty^{\theta,\alpha}(x) = 1 \text{ for all } x \in A) = \int_0^1 s^{|A|} dF_{\theta,\alpha}(s).$$

To identify the mixture $dF_{\theta,\alpha}$, we call upon the "mystery guest in the scenario for critical clustering": the Fisher–Wright diffusion Y_t; that is, the strong Markov process on $[0,1]$ with generator

$$Lf(x) = \tfrac{1}{2} x(1-x)f''(x) \quad 0 < x < 1.$$

It is well known in the mathematical genetics literature that Y_t and D_t satisfy the "duality equation"

(14) $$E_\theta(Y_t^n) = E_n(\theta^{D(t)}).$$

(For a proof see Tavaré (1984).) If we let $q_t(\theta,\cdot) = P_\theta(Y_t \in \cdot)$ be the transition probability for Y_t, then we can put the pieces together to compute $F_{\theta,\alpha}$.

$$P(\xi_\infty^{\theta,\alpha}(x) = 1 \text{ for } x \in A) = \sum_{k=1}^{n} \theta^k p_{n,k}(\alpha)$$

$$= E_n \, \theta^{D(\log(1/\alpha))} = \int_0^1 s^n \, q_{\log(1/\alpha)}(\theta,ds)$$

that is, $F_{\theta,\alpha}$ is just the distribution at time $\log(1/\alpha)$ of the Fisher–Wright diffusion starting from θ. The last result is an unexpectedly tidy solution to the problem of identifying $F_{\theta,\alpha}$. It and (12) are the beautiful answers referred to at the beginning of the section. Even though no one in their right mind could see this result in a computer simulation, we close this section by showing some pictures of the two dimensional voter model.

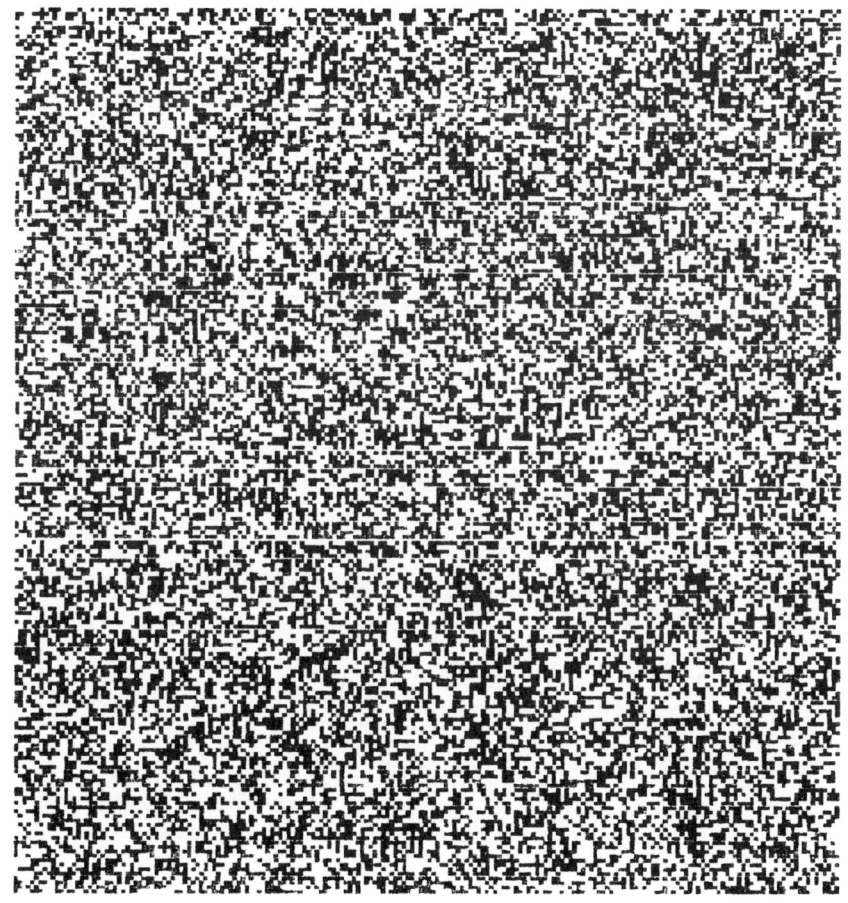

TIME 0

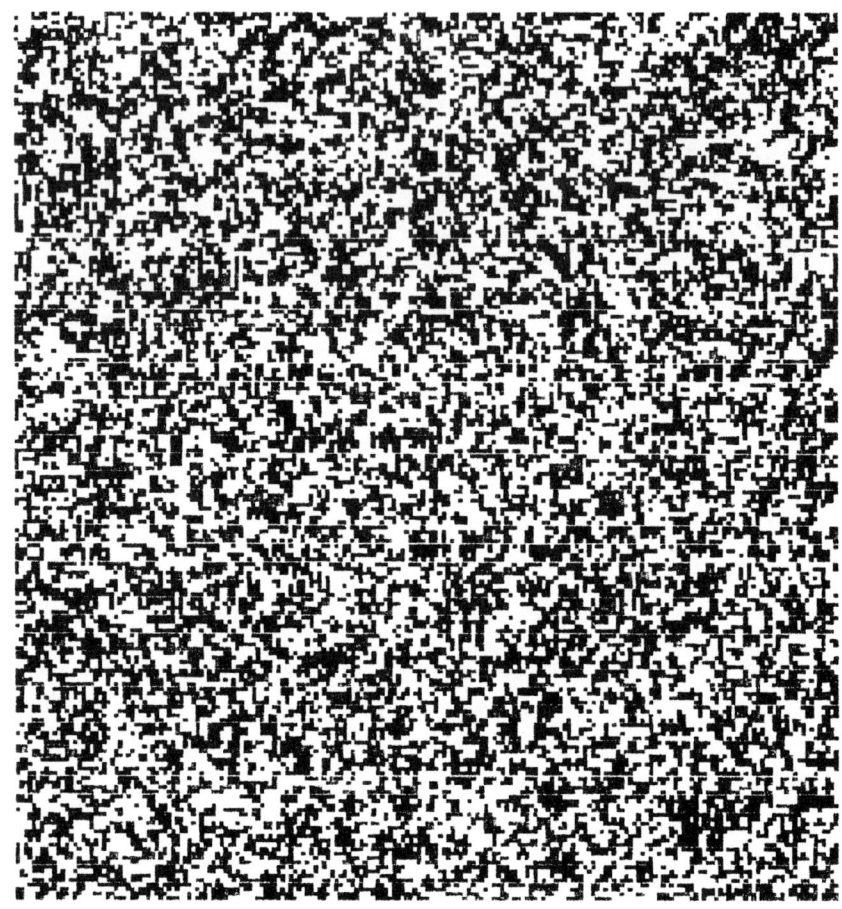

TIME 1

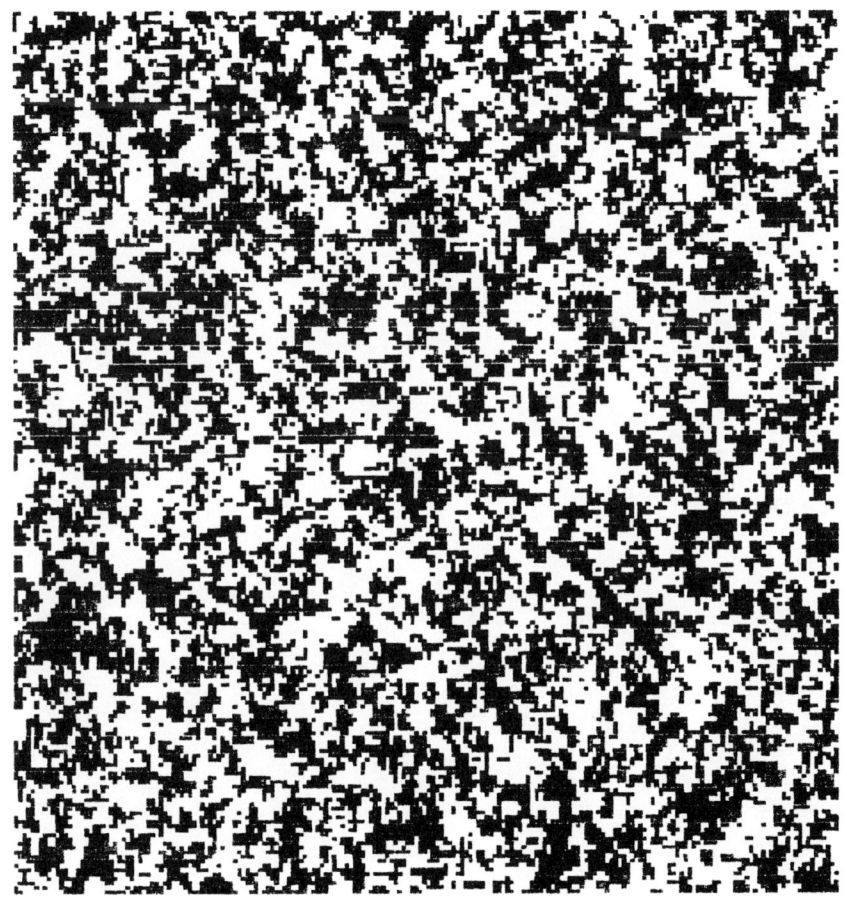

TIME 5

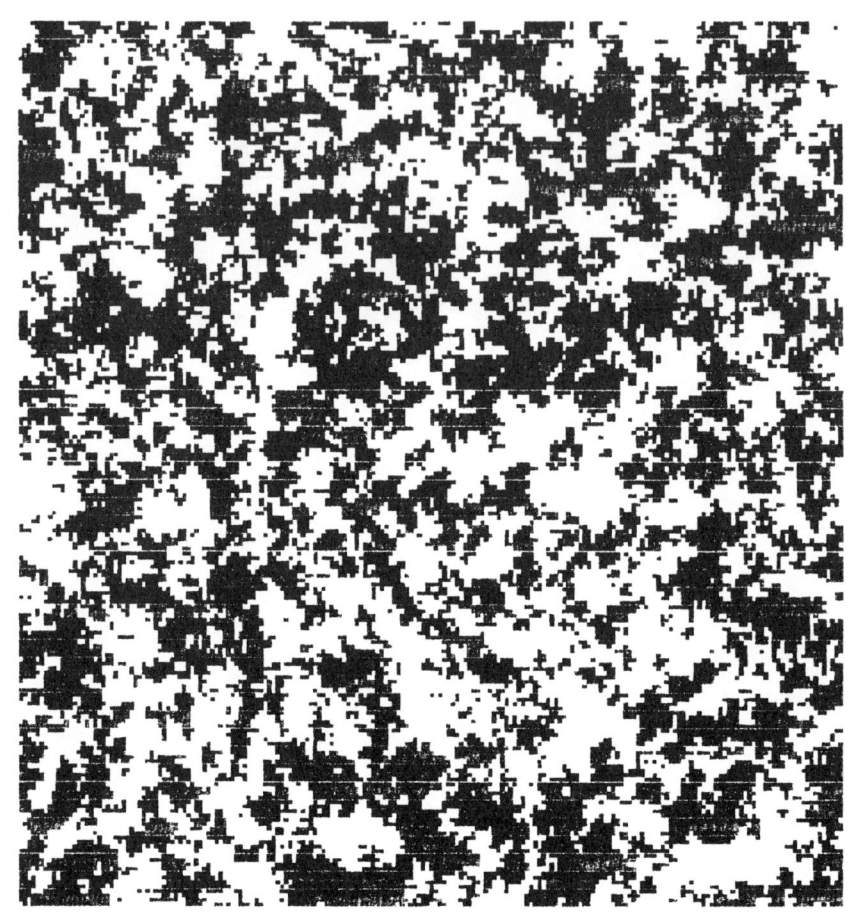

TIME 25

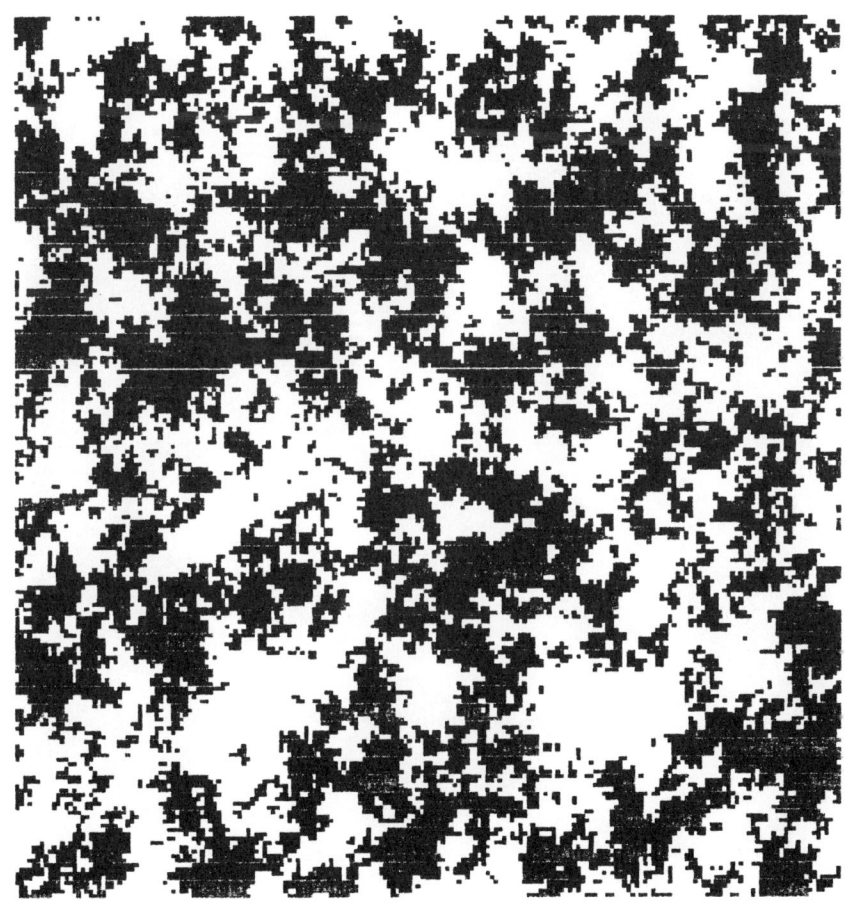

TIME 100

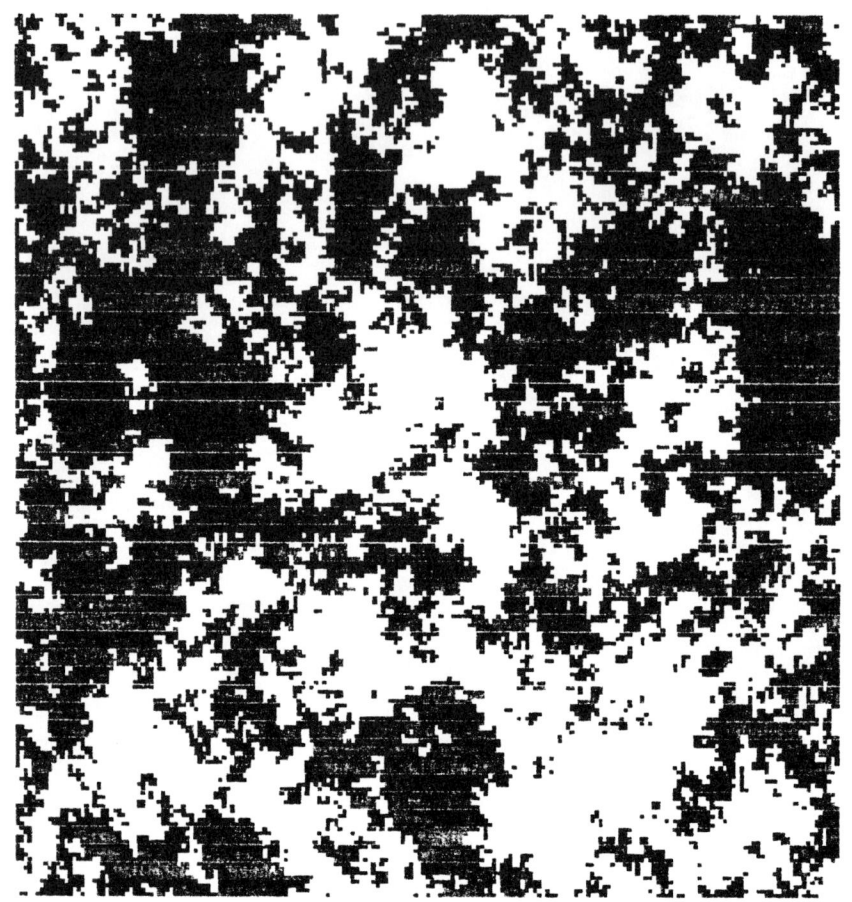

TIME 400

10b Voter Model on a Finite Set

In this section, we consider what happens when we consider the voter model on the d dimensional torus $\pi_N = (\mathbb{Z} \bmod N)^d$. Since we are dealing with a Markov chain on a finite set that has absorbing states, the asymptotic behavior is trivial. At some time τ_N, all the opinions become the same, and nothing changes after that time. Here we are interested in how τ_N grows as $N \to \infty$. To keep things simple, we will assume that initially the sites are all different colors. (If you have trouble thinking of that many colors, think of the initial type of each site as being indicated by its position in $\mathbb{Z}^d$.) The reason for considering this case will become clear when we begin the proofs. At the end of the section, we describe what happens starting from a product measure.

The main result of this section is

(1) **Theorem.** As $N \to \infty$

$$
\begin{array}{llll}
\tau_N / N^2 & \Rightarrow & F_1 & \text{in } d = 1. \\
\tau_N / N^2 \log N & \Rightarrow & F_2 & \text{in } d = 2. \\
\tau_N / N^d & \Rightarrow & F_d & \text{in } d \geq 3.
\end{array}
$$

Here $\Rightarrow$ denotes convergence in distribution.

We will content ourselves to derive (1) rather than prove it; that is, we will explain why the normalizing constants are what they are, and we will compute the limiting distributions. In doing this, we will use ideas we learned from a course taught by David

Aldous at Cornell in the fall of 1986. He is not to be blamed for the holes we have left

in our proof. Detailed proofs (by a different method) can be found in Cox (1988).

Proof: Since we have assumed that initially all the sites are different colors, duality

implies that τ_N has the same distribution as the time it takes coalescing random walks

starting from all points of π_N to be reduced to one particle, a time we will also call τ_N.

Hereafter, we will deal only with coalescing random walks, so the duplicated notation

should not cause any confusion.

We begin with the case d=1. In this case, as David Aldous first observed, τ_N

can be identified as the first time two initially adjacent particles "hit the hard way".

To explain the phrase in quotation marks, consider the coalescing random walk as

taking place on $\mathbb{Z}$ starting with all sites occupied, and with particles that differ by a

multiple of N moving in parallel. Let X_t^k be the position at time t of the particle that is

initially at k. If $X_t^{j+1} - X_t^j$ becomes N, then we say that j and j+1 have "hit the hard

way." It is easy to see that when this occurs, all the particles must have coalesced to 1.

(See Figure 10.3.)

If we let $\sigma_j = \inf\{\ t : X_t^{j+1} - X_t^j = N\ \}$, then results in the last paragraph show

that

$$\tau_N \leq \min_{0 \leq j \leq N-1} \sigma_j.$$

To prove there is equality here, consider the situation where there are just two

particles left. One of the particles, call him Fred, consists of the particles originally at

a≤j≤b where 0≤a≤b≤N−1, and the other one, call her Ethel, consists of the rest. If just

before the final collision at time τ_N,

$$X^{Fred}(\tau_N-) = X^{Ethel}(\tau_N-)-1$$

(where τ_N- indicates the limit as t ↑ τ_N), then a and a–1 have "hit the hard way." Otherwise b and b+1 did. (See Figure 10.3 where the first case is drawn.)

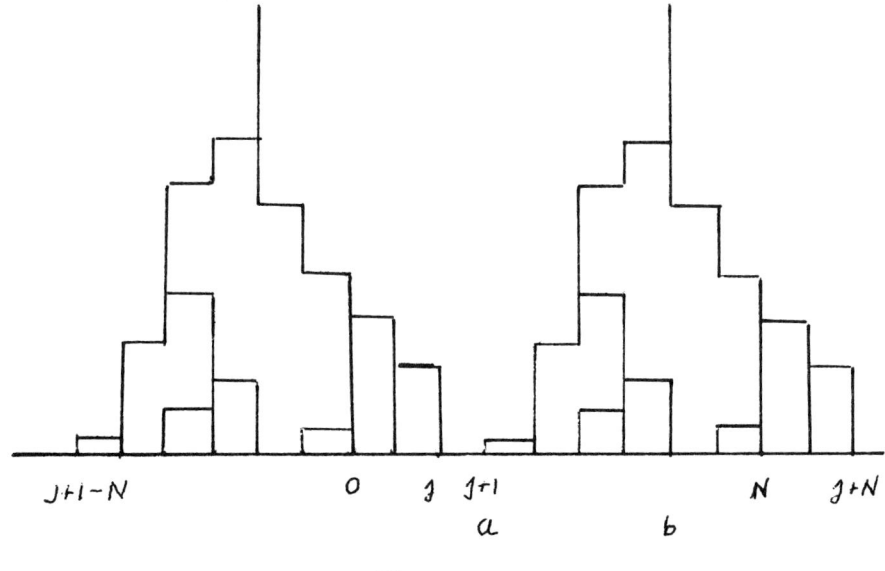

Figure 10.3

At this point we have shown

$$\tau_N = \min_{0 \leq j \leq N-1} \sigma_j.$$

A little more thought reveals that the minimum is attained for only one j, and all the other σ's are +∞. If we let J be the index with $\tau_N = \sigma_J$, then obvious symmetry implies P(J = j) = 1/N, and P($\tau_N \geq t$ | J = j) = P($\tau_N \geq t$). The last quantity is easy to compute. Suppose we condition on J = 0, then since $X_t^1 - X_t^0$ is a simple random walk that makes jumps at rate 2, what we want to do is compute the distribution of the time it takes to hit N given that it hits N before it hits 0.

To avoid headaches when comparing with known formulas for simple random walk, let R_t be a continuous time simple random walk that makes jumps at rate 1, and let S_m be the embedded discrete time chain (i.e. S_m is the position after the mth jump). If we condition these processes to hit N before 0, stop them when they hit N, and let R_t^N, S_m^N be the processes that result, then S_m^N is a Markov chain on {1, 2, ..., N } with an absorbing state at N and transition probability given by

$$P(\, S_{m+1}^N = x+1 \mid S_m^N = x \,) = \frac{\frac{1}{2} \cdot \frac{x+1}{N}}{\frac{x}{N}} = (x+1)/2x$$

for $0 < x < N$ (since $P_x(\, S_m$ hits N before 0 $) = x/N$), and R_t^N is the continuous time chain that makes jumps at rate one and has embedded chain S_m^N.

If you are familiar with what happens when random walks are conditioned, then you will immediately see that

(2) $$R^N(N^2t)/N \Rightarrow B_3(t \wedge H_1)$$

where $B_3(t)$ is a three dimensional Bessel process started at 0, and $H_1 = \inf\{t: B_3(t) = 1 \}$ is the time the process first hits 1. If you have not heard these things before, then the conclusion can be made plausible by noting that if $X^N(t) = R^N(N^2t)/N$, then

$$\frac{d}{dt} E(\, X^N(t){-}x \mid X^N(0) = x \,) = N^2 \left[\frac{Nx+1}{2Nx}{\cdot}\frac{1}{N} + \frac{Nx-1}{2Nx}{\cdot}\frac{-1}{N} \right] = 1/x,$$

and

$$\frac{d}{dt} E(\, (X^N(t){-}x)^2 \mid X^N(0) = x \,) = N^2 \left[\frac{Nx+1}{2Nx}{\cdot}(\tfrac{1}{N})^2 + \frac{Nx-1}{2Nx}{\cdot}(\tfrac{1}{N})^2 \right] = 1.$$

So the infinitesimal mean and variance of X^N converge to those of the diffusion with generator

$$L = \frac{1}{2} \cdot \frac{d^2}{dx} f + \frac{1}{x} \cdot \frac{d}{dx} f$$

that is, the three dimensional Bessel process. (Recall that the d dimensional Bessel process has drift $(d-1)/2x$.)

While the calculations above give the answer quickly, one can prove the result without using anything fancy, so we will. Let $T_N = \inf\{ t : R_t^N = N \}$. By considering what happens at the first jump, we get for $1 \leq x < N$

$$E_x \exp(-\theta\, T_N) = \frac{1}{\theta+1} \cdot \left[\frac{x+1}{2x} E_{x+1} \exp(-\theta\, T_N) + \frac{x-1}{2x} E_{x-1} \exp(-\theta\, T_N) \right]$$

(the first factor being the Laplace transform of the time to the first jump). Letting $\Psi(x) = x E_x \exp(-\theta\, T_N)$ and rearranging gives

(3) $$\Psi(x+1) - 2(\theta+1)\Psi(x) + \Psi(x-1) = 0,$$

an equation that holds for $1 \leq x < N$ if we take $\Psi(0) = 0$.

It is well known (and easy to check if you don't) that all solutions of (3) are of the form $C_1 r_1^x + C_2 r_2^x$, where $r_1 > r_2$ are the solutions of $r^2 - (2\theta+2)r + 1 = 0$, that is

$$r_i = \frac{(2\theta+2) \pm ((2\theta+2)^2 - 4)^{1/2}}{2} = (\theta+1) \pm (\theta^2 + 2\theta)^{1/2}.$$

The $\Psi(x)$ we want has $\Psi(0) = 0$ and $\Psi(N) = N$, so $C_1 = -C_2$. A little fiddling gives

$$\Psi(x) = N \left[\frac{r_1^x - r_2^x}{r_1^N - r_2^N} \right],$$

and recalling the definition of $\Psi(x)$, we have

$$E_1 \exp(-\theta\, T_N) = \left[\frac{r_1 - r_2}{r_1^N - r_2^N} \right].$$

(To reconcile this with Cox's formula, observe that the quadratic equation can be written as $r-(2\theta+2)+1/r = 0$ so $r_2 = 1/r_1$.) Putting $\theta = \lambda/N^2$, we have

$$r_1 = 1 + \frac{\lambda}{N^2} + \left[\frac{2\lambda}{N^2} + \frac{\lambda}{N^4} \right] = 1 + \frac{\sqrt{2\lambda}}{N} + O(N^{-2}),$$

$$r_2 = 1 + \frac{\lambda}{N^2} - \left[\frac{2\lambda}{N^2} + \frac{\lambda}{N^4} \right] = 1 - \frac{\sqrt{2\lambda}}{N} + O(N^{-2}).$$

So as $N \to \infty$

$$N(r_1 - r_2) \to 2\sqrt{2\lambda}, \qquad r_1^N - r_2^N \to 2\sinh(\sqrt{2\lambda}),$$

and recalling τ_N has the same distribution as $T_N/2$,

(4) $$E_1 \exp(-\alpha\tau_N/N^2) = E_1 \exp(-\alpha T_N/2N^2) \to \sqrt{\alpha}/\sinh(\sqrt{\alpha}).$$

We leave it to the reader to check that this is the same as the formula of Ciesielsky and Taylor for the exit time from the unit ball for three dimensional Brownian motion. (See Theorem 4.2.20 on p.88 of Knight (1981).)

The reasoning above clearly breaks down in $d > 1$. To derive (1) in $d \geq 2$, we will use a special argument which fails in $d < 2$. (Fortunately, there are no integers in between 1 and 2!) We begin with $d \geq 3$, since it is easier than the borderline case $d = 2$. We start our analysis of that case by considering how long it takes two independent random walks X_t and Y_t, starting from randomly chosen points on the torus $\pi_N = (Z \bmod N)^d$, to hit. Let $T_N = \inf\{t : X_t = Y_t\}$. It is easy to see that for any $M > 0$

$$(5) \qquad P(\, T_N \le t \,) \le (t+M)/(N^d \int_0^M P(\, S_r = 0 \,)\, dr\,),$$

where S_r is a simple random walk on Z^d that makes transitions at rate 2.

Proof: At any time s, X_s and Y_s are independent and uniformly distributed on π_N so

$$(t+M)/N^d = E|\{\, s \le t+M : X_s = Y_s \,\}|$$

$$\ge E\left[\int_{T(N)}^{T(N)+M} 1_{(X_s = Y_s)}\, ds \; ; T_N \le t \right]$$

$$\ge P(\, T_N \le t \,) \int_0^M P(\, S_r = 0 \,)\, dr.$$

If we let $g = \int_0^\infty P(S_r = 0\,)\, dr$, $t = \theta N^d$, and let $M \to \infty$ with $M/N^d \to 0$, then we get

$$(6) \qquad \limsup_{N \to \infty} P(\, T_N \le \theta N^d \,) \le \theta/g.$$

The last result is crude, but is almost the right answer. We will show that

$$P(\, T_N \le \theta N^d \,) \to 1 - e^{-\theta/g}.$$

The intuition behind the appearance of the exponential distribution is simple. (6) tells us that the time required for two particles to hit is at least ϵN^d, which is much larger than the N^2 time scale on which a random walk on π_N reaches equilibrium. Therefore, if we condition on $T_N > N^d$ and let K be large then at time $N^d + KN^2$ the locations of the two particles are almost independent and uniformly distributed on π_N. Letting $N,K \to \infty$ and $K/N^{d-2} \to 0$, we see

$$\lim_{N \to \infty} P(T_N > N^d(t+s) \mid T_N > N^d t) = \lim_{N \to \infty} P(T_N > N^d s),$$

that is, the limit of T_N/N^d has the lack of memory property that characterizes the exponential distribution.

We will begin the process of turning the intuition above into a proof by making our statement about convergence to equilibrium precise:

(7) **Lemma.** Let S_t^N be a random walk on π_N that makes jumps at rate 2. There is a constant $\epsilon > 0$ (which depends upon the dimension but is independent of N) so that for all integers k

$$\frac{1}{2} \sum_{y \in \pi_N} |P(S^N(kN^2) = y) - 1/N^d| \le (1-\epsilon)^k.$$

Proof: The local central limit theorem implies that if $N \to \infty$ and $y_N/N \to y$ then

$$N^d P_0(S(N^2) = y_N) \to \varphi(y)$$

where φ is the d dimensional Normal density with mean 0 and covariance $2I/d$, and the absence of the superscript indicates we are dealing with random walk on $\mathbb{Z}^d$. The last observation implies there is an $\epsilon > 0$ so that

$$P_0(\ S^N(N^2)=y\ ) \geq \epsilon/N^d \text{ for all N and } y \in \pi_N.$$

Since the sum in (7) = the sum of the negative parts of the difference = the total variation distance between the two measures, the result holds for $k = 1$.

To prove the result for $k \geq 2$ we iterate. Let $q(x,y) = P_x(S^N(N^2) = y\ )$, $\mu(y) = 1/N^d$, and $q^\epsilon(x,y) = (1-\epsilon)^{-1}(q(x,y) - \epsilon\mu(y))$. $q^\epsilon(x,y)$ is a transition probability and

$$q(x,y) = (1-\epsilon)\ q^\epsilon(x,y) + \epsilon\ \mu(y).$$

Iterating and using the fact that $\mu(y)$ is a stationary distribution for $q(x,y)$ (and hence for $q^\epsilon(x,y)$) gives

(8) $$q_k(x,y) = (1-\epsilon)^k q_k^\epsilon(x,y) + (1-(1-\epsilon)^k)\mu(y),$$

where the subscript k indicates the kth iterate. We will have more to say about the derivation of (8) in a minute. Before getting sidetracked, we would like to observe that (8) and the observation at the end of the last paragraph imply the desired result.

It is not hard to verify (8) by induction. One writes

$$q_k(x,y) = \sum_z [(1-\epsilon)\ q^\epsilon(x,z) + \epsilon\ \mu(z)]\ q_{k-1}(z,y),$$

uses the induction hypothesis on $q_{k-1}(z,y)$, and 3–10 minutes later you are done. For arguments below, it is useful to notice that it can be done probabilistically also. Let $U_1, U_2, \ldots$ and $V_1, V_2, \ldots$ be i.i.d. sequences with U_i uniform on π_N, and $P(\ V_i = 1\ )$

$= \epsilon$, $P(V_i = 0) = 1 - \epsilon$. If we let $X_k = S^N(kN^2)$, then we generate X_{k+1} by a two step procedure. First we "flip a coin" with probability ϵ of heads. If a heads appears (i.e., $V_{k+1} = 1$), we let $X_{k+1} = U_{k+1}$. If a tails appears (i.e., $V_{k+1} = 0$) and $X_k = x$, we put X_{k+1} at a location with distribution $q^\epsilon(x, \cdot)$, but which is otherwise independent of X_0, ..., X_k. It should be clear that when the first heads appears the process "forgets its initial state and reaches equilibrium". If $X_0 = x$ and X makes k transitions without a heads, an event with probability $(1-\epsilon)^k$, then the location of X_k has distribution $q_k^\epsilon(x, \cdot)$. If a heads appears in the first k tosses, an event of probability $1 - (1-\epsilon)^k$, then X_k is uniformly distributed on π_N, and we have the expression in (8). The last construction is useful for proving results about Markov chains on general state spaces. (See Athreya and Ney (1978).)

To prove our limit theorem for T_N / N^d, we divide time up into big blocks with little blocks in between. Let

$$I_m = [(m-1)N^\alpha, mN^\alpha - N^\beta], \qquad \text{and} \qquad J_m = [mN^\alpha - N^\beta, mN^\alpha],$$

where $2 < \beta < \alpha < d$. Let γ be such that $\beta < \gamma < \alpha$. Since $P(X_t = Y_t) = 1/N^d$, it is easy to see by using (5) with $M=1$ that

$$P(X_t = Y_t \text{ for some } t \in J_m \text{ with } m \le N^{d-\gamma}) \to 0.$$

So the probability X_t and Y_t hit in one of the little blocks at some $t \le N^{d-\gamma+\alpha} \to 0$.

As for the big blocks, it is easy to see that the argument for (5) can be reversed to give

(9) $P(\ X_t = Y_t \text{ for some } t \in I_m\ ) \geq (N^{\alpha}-N^{\beta})/(N^d \int_0^{N^{\alpha}} P_0(S_r^N = 0)\ dr),$

where S_t^N is a simple random walk on π_N that makes jumps at rate 2.

Proof: It suffices to prove the result when m = 1.

$$(N^{\alpha}-N^{\beta})/N^d = E|\{\ s \leq N^{\alpha}-N^{\beta} : X_s = Y_s\ \}|$$

$$\leq E\left[\ \int_{T(N)}^{T(N)+N^{\alpha}} 1_{(X_s = Y_s)}\ ds\ ;\ T_N \leq N^{\alpha}-N^{\beta}\ \right]$$

$$\leq P(\ T_N \leq N^{\alpha}-N^{\beta}\ ) \int_0^{N^{\alpha}} P_0(\ S_r^N = 0\ )\ dr.$$

Now I claim, and leave for you to prove, that

(10) $$\int_0^{N^{\alpha}} P_0(\ S_t^N = 0\ )\ dt = \int_0^{N^{\alpha}} P_0(\ S_t = 0\ )\ dt + o(1)$$

Sketch: The convergence to equilibrium in (7) takes care of times between $N^{2+\epsilon}$ and N^{α}. (Here we use the fact that $P(\ S_t^N = 0\ ) \approx N^{-d}$ and $\alpha < d$.) Times $\leq N^{2-\epsilon}$ can clearly be neglected because the probability of $\|S_t\|_{\infty} \geq N$ is very small. As for times in between, observe that the local limit theorem implies that for $z \in \mathbb{Z}^d$,

$$P(\ S_t = Nz\ ) \approx \varphi(Nz/\sqrt{t})/t^{d/2}.$$

Since $t \leq N^{2+\epsilon}$, we only have to worry about z with $\|z\|_{\infty} \leq N^{\epsilon}$. Replacing $\varphi(Nz/\sqrt{t})$ by $\varphi(0)$ and t by $N^{2-\epsilon}$, summing over $\|z\|_{\infty} \leq N^{\epsilon}$, and integrating over $N^{2-\epsilon} \leq t \leq N^{2+\epsilon}$, gives a result that is o(1) if ϵ is small enough. We leave to the reader the thankless job of turning the last sketch into a real proof.

Combining (9) and (10) with the lower bound in (5), shows

$$P(\ X_t = Y_t \text{ for some } t \in I_m\ ) \sim N^{\alpha - d}/g,$$

where

$$g = \int_0^\infty P_0(\ S_t = 0\ )\ dt$$

is the expected amount of time a simple random walk in $\mathbb{Z}^d$ that moves at rate 2 stays at 0. If we let η_m be random variables that indicate whether or not a hit occurred during I_m, then using the probabilistic proof of (8) it is easy to show that we can construct independent random variables ζ_m with the same distribution as that of the η_m's so that

$$P(\ \eta_m \neq \zeta_m \text{ for some } m \leq tN^{d-\alpha}\ ) \leq tN^{d-\alpha}(1-\epsilon)^{N^{\beta-2}} \to 0.$$

Since

$$P(\ \zeta_m = 0 \text{ for all } m \leq tN^{d-\alpha}\ ) \approx (1 - N^{\alpha-d}/g)^{tN^{d-\alpha}} \to e^{-t/g},$$

as $N \to \infty$, it follows that we have

(11) $$P(\ T_N \geq tN^d\ ) \to e^{-t/g},$$

the result we promised to prove.

Having studied the behavior of T_N in detail, the next step is to consider what happens when we start three random walks X_t, Y_t, and Z_t from randomly chosen points in π_N, and look at $T_N^3 = $ the first time two of them collide. Using the same blocks as before, and observing that $P(\, X_t = Y_t = Z_t\,) = 1/N^{2d}$, it is easy to see that

$$P(\text{ a collision occurs at some } t \in I_m \,) \sim \begin{bmatrix} 3 \\ 2 \end{bmatrix} N^{\alpha-d}/g,$$

and the events for different m are asymptotically independent, so

$$P(\, T_N^3/N^d \geq t\,) \to \exp(-\begin{bmatrix} 3 \\ 2 \end{bmatrix} t/g).$$

The argument above remains valid if 3 is replaced by k, so putting these results together with a small leap of faith gives

(12) $$P(\, T_N/N^d \leq t\,) \to P(\, \sum_{k=2}^{\infty} W_k / \begin{bmatrix} k \\ 2 \end{bmatrix} \leq t\,)$$

where $W_2, W_3, \ldots$ are i.i.d. with $P(\, W_i > t\,) = e^{-t/g}$. The intuition behind the result should be clear. If we let τ_N^k be the first time the coalescing random walk has $\leq$ k particles, and IF the particles are far enough apart so that the probability of a collision in the next $N^{2+\epsilon}$ units of time can be ignored, then at time $\tau_N^k + N^{2+\epsilon}$ the locations of the k particles will be very close to that of k independent uniformly distributed particles. This means that $\tau_N^{k-1} - \tau_N^k$ will have a distribution close to that of $N^d W_k / \begin{bmatrix} k \\ 2 \end{bmatrix}$, and be almost independent of τ_N^k.

The IF in the last paragraph is not too hard to justify, but requires more work than we are willing to do, so we will just sketch a proof and leave the details to the reader. The probability (in equilibrium) of having three particles within a distance N^{δ} is at most $CN^{2(\delta-d)}$, so if $\delta < d/2$ the probability this will happen at some time $\leq$

$N^{d+\epsilon}$ is small (if ϵ is small enough). On the other hand, the probability two random walks starting at x and y in Z^d will ever hit is $\approx \|x-y\|_2^{d-2}$ (the answer for Brownian motion). The last observation allows us to estimate the probability of a collision at a time $\leq N^{2-\epsilon}$. Estimates similar to ones we did not give in the proof of (10) control the times between $N^{2-\epsilon}$ and $N^{2+\epsilon}$, and the rest is left to the reader.

The last paragraph is, admittedly, rather far from a proof. For complete details using the methods of the last section, see Cox (1988). He also demonstrates the other detail that is missing from our "proof" of (12). The limit of τ_N/N^d is the same as the limit as $k \to \infty$ of the limits of $(\tau_N - \tau_N^k)/N^d$.

Having gone into a lot of detail about the proof in $d \geq 3$, we will be brief in our treatment of $d = 2$. (5), whose proof did not rely on the fact that $d \geq 3$, generalizes immediately to

$$(13) \qquad P(T_N \leq t) \leq (t+M)/(N^2 \int_0^M P_0(S_r = 0) \, dr).$$

Since S_r makes transitions at rate 2, and each component of the jump distribution has variance $1/2$,

$$\int_0^M P_0(S_r = 0) \, dr \sim \int_1^M 1/(2\pi r) \, dr = \frac{1}{2\pi} \log M.$$

Letting $t = \theta N^2 \log N$, $M = N^2$, and $N \to \infty$ gives

$$\limsup_{N \to \infty} P (T_N \leq \theta N^2 \log N) \leq \pi \theta.$$

The times we are looking at are about $N^2 \log N$, and the convergence to equilibrium takes place on an N^2 scale, so this time we let the large blocks I_m have length $N^2(\log N)^{2/3}$ and the small blocks have length $N^2(\log N)^{1/4}$. Taking $M = N^2$ in (13) shows

$$P (X_t = Y_t \text{ for some } t \in J_m \text{ with } m \leq (\log N)^{1/2})$$
$$\leq (\log N)^{1/2} \, 2(\log N)^{1/4}/(C \log N) \to 0.$$

Using (13) and an inequality analogous to (9), shows

$$P(X_t = Y_t \text{ for some } t \in I_m) \sim \pi/(\log N)^{1/3}.$$

Let η_m be random variables that indicate whether a hit occurred during I_m. Using the probabilistic proof of (8), it is easy to show that we can construct independent random variables ζ_m with the same distribution as the η_m's so that

$$P(\eta_m \neq \zeta_m \text{ for some } m \leq (\log N)^{1/2}) \leq (\log N)^{1/2}(1-\epsilon)^{(\log N)^{1/4}} \to 0.$$

Since

$$P(\zeta_m = 0 \text{ for all } m \leq t(\log N)^{1/3}) \approx (1- \pi/(\log N)^{1/3})^{t(\log N)^{1/3}} \to e^{-\pi t}$$

it follows that

(14) $$P (T_N \geq tN^2 \log N) \to e^{-\pi t}.$$

Having found the limiting behavior of T_N, the heuristic we used above takes over to tell us the limiting behavior of τ_N

(15) $$P(\ \tau_N/(N^2 \log N) \le t\ ) \to P(\ \sum_{k=2}^{\infty} W_k/\begin{bmatrix} k \\ 2 \end{bmatrix} \le t\ )$$

where $W_2, W_3, \ldots$ are i.i.d. with $P(\ W_i > t\ ) = e^{-\pi t}$. The proof of (15) is similar to that of (12), but is more delicate. Again, the reader can find the details in Cox (1988).

We will devote the rest of this section to two extensions of (1). We begin by computing the asymptotic behavior of $E\tau_N$, and start with the case d=2 because the formula is handy. Taking expected values in (15) and ignoring the justification of this step gives

$$E\tau_N/(N^2 \log N) \to \sum_{k=2}^{\infty} EW_k/\begin{bmatrix} k \\ 2 \end{bmatrix} = \frac{1}{\pi} \cdot \sum_{k=2}^{\infty} \frac{2}{k(k-1)} = \frac{2}{\pi},$$

since the sum telescopes. The limit in $d \ge 3$ is the same, except for the fact that we divide by N^d and $EW_k = g$.

To determine the asymptotic behavior of $E\tau_N$ in $d = 1$, we observe that (4) gives the asymptotic behavior of the Laplace transform of τ_N/N^2. To find the limit of $E\tau_N/N^2$ we differentiate $\sqrt{\alpha}/\sinh(\sqrt{\alpha})$ and set $\alpha = 0$, or what is easier, observe

$$\sinh(x) = x + \frac{x^3}{3!} + O(x^5),$$

so

$$\sqrt{\alpha}/\sinh(\sqrt{\alpha}) = 1 - \alpha/3! + O(\alpha^2).$$

Interchanging the limits $N \to \infty$ and $\alpha \to 0$, it follows that $E\tau_N/N^2 \to 1/6$.

A second, somewhat simpler, route to the answer is to observe that (2) tells us that $T_N/N^2 \to H_1 = \inf\{\ t : B_3(t) = 1\ \}$, where $B_3(t)$ is a three dimensional Bessel process started at 0, and $\tau_N \overset{d}{=} T_N/2$. To compute EH_1, observe that if W_t is a three dimensional Brownian motion, then $B_3(t) \overset{d}{=} \|\ W_t\ \|_2$ and $\|\ W_t\ \|_2^2 - 3t$ is a martingale. The optional stopping theorem implies $0 = 1 - 3EH_1$, so $EH_1 = 1/3$, and again we have $E\tau_N/N^2 \to 1/6$.

Putting the pieces together, and letting $G =$ the expected time at 0 for simple random walk in $\mathbb{Z}^d$ that makes jumps at rate one $(= 2g)$ gives:

(16) Theorem. As $N \to \infty$

$$E\tau_N \ \sim \ \begin{array}{ll} (1/6)\ N^2 & \text{in } d = 1 \\ (2/\pi)\ N^2 \log N & \text{in } d = 2 \\ G\ N^d & \text{in } d \geq 3. \end{array}$$

Finally, we come to the question of the asymptotic behavior of τ_N^θ, the time to reach consensus when the initial state is a product measure with m colors in which the ith has density θ_i. In $d \geq 2$ only one new ingredient is needed: if there are k particles in the coalescing random walk, then the probability a consensus exists in the voter model is $\Sigma_i\ \theta_i^k$. Adding this to the picture of the τ_N^k that emerged in the proofs of (12) and (15), we can compute the asymptotic behavior of τ_N^θ. We will only mention the result for the expected value.

(17) Theorem. In $d \geq 2$, $E\tau_N^\theta \sim -(\ \Sigma_i\ (1-\theta_i)\ \log(1-\theta_i)\ )\ E\tau_N$ as $N \to \infty$.

Proof: Since the probability a consensus has been reached when there are k particles is

$\Sigma_i \; \theta_i^k$, Fubini's theorem implies that

$$E\tau_N^\theta \sim c_N \cdot \sum_{k=2}^{\infty} (1- \Sigma_i \; \theta_i^k) EW_k / \begin{bmatrix} k \\ 2 \end{bmatrix},$$

where $c_N = N^2 \log N$ in d=2 and $c_N = N^d$ in d≥3. To evaluate the sum on the right we observe:

$$\sum_{k=0}^{\infty} \theta^k = 1/(1-\theta),$$

$$\sum_{k=0}^{\infty} \theta^{k+1}/(k+1) = -\log(1-\theta),$$

$$\sum_{k=2}^{\infty} \theta^k \left[\frac{1}{k-1} - \frac{1}{k} \right] = -\theta\log(1-\theta) + \theta + \log(1-\theta).$$

So

$$E\tau_N^\theta \sim E\tau_N \left[1 - \Sigma_i \; [\; \theta_i + (1-\theta_i)\log(1-\theta_i) \;] \right],$$

and recalling $\Sigma_i \; \theta_i = 1$, gives the result quoted above.

When $\theta_i = 1/c$ for $1 \leq i \leq c$, that is, we start with c colors that have equal probability, then (17) becomes

(18) $$E\tau_N^\theta \sim -(c-1)\log(1-c^{-1}) \; E\tau_N.$$

Taking c = 4, d = 2, and N = 25 (these are the values in the program D2VOTER4 on IPSmovies, see the advertisement in the Introduction), we have

$$E\tau_N \approx (2/\pi)\, 625 \log(25) = 1280$$
$$E\tau_N^{\theta} \approx 3 \log(3/4)\, E\tau_N = 1105.$$

To check the last result, and to see whether the approximation is valid for N = 25, we simulated coalescing random walk 20 times and found the following values of τ_N:

561,   639,   711,   782,   840,   977,   980,   1073,   1086,   1129,

1247,   1255,   1280,   1586,   1685,   1685,   1740,   1858,   1890,   3284

(yes 1685 occured twice!). The average of these 20 observations is 1314.4, which is in reasonable agreement with the values given above. Running the voter model starting from 4 colors, 10 times gave an average value of $\tau_N^{\theta} = 1011.4$, again consistent if somewhat low. For readers who doubt the usefulness of checking theorems by computer simulation, we would like to point out the simulations above revealed that Cox's original constants were off by a factor of 4.

Having praised the computer, it is only fair to say that it does not always give the right answer. (One of my colleagues, who shall remain na $\mathcal{MB}$ less, claims that the computer is wrong two–thirds of the time.) Simulating coalescing random walk on $\mathbb{Z}$ mod 198 starting from all sites occupied (using D1VOTER198 on IPSmovies) twenty–five times gave an average value of 13,178 for τ_N. Since $(198)^2/6 = 6534$, I wasted a lot of time checking and rechecking the proof to see where I had lost a factor of 2, before I figured out that the random number generator was at fault.

In D1VOTER198, a statement "if Random(2) = 0 then..." is used to determine whether the random walk will step right or left. According to the Turbo Pascal manual, "Random(Num) returns a random number greater than or equal to 0 and less than Num. Num and the random number are both integers." Being positive integers, they are restricted to the range 0 to 32767, and the random number generator has a period $\leq$ 32767. (A little sleuthing shows = holds.) The simulations above use a significant fraction of the period, and one runs into what I call the "Brownian Bridge Phenomenon." The sum of 1−2*random(2) over the period is 0, so if n = 32767, and S_m is the sum of the first m numbers, $S_{[nt]}/\sqrt{n}$ is close to Brownian bridge, not the Brownian motion we want.

I will leave it to the reader to decide for herself, if this explains the problem encountered above. Rewriting the statement in question as "if Random < 0.5 then..." (Random $\in$ [0,1) has 2^{32} possible values) gets rid of the problem. The average value of τ_N in 50 simulations was 6469, in beautiful agreement with theory. The last sentence is also a beautiful example of lying with statistics. My first sample of size 25 had a mean of 7768, so I ran the program 25 more times. The second 25 observations had a mean of 5170, so I combined the two and stopped.

The means of the two samples of size 25 are about as close to the true value as we should expect. Replacing the random walk by Brownian motion, and using the fact that $B_t^4 - 6B_t^2 t + 3t^2$ is a martingale, one finds that the exit time T_a from (−a,a) has

$$3ET_a^2 = 6a^2 \, ET_a - a^4 = 5a^4,$$

the variance $\sigma^2(T_a) = 2a^4/3$, and the standard deviation, $\sigma(T_{98}/2) \approx 1600$. (We divide by 2 because the random walk we are interested in moves at rate 2.) By finding constants c_n to make $S_n^4 - 6S_n^2 n + c_n$ a martingale, the last computation can be done for simple random walk. We leave this as an exercise for the reader because it would not improve our estimate of $\sigma(\tau_N)$ by very much.

11 The Contact Process, II

11a Renormalized Bond Construction

In this section, we introduce a construction that allows us to reduce questions about the supercritical contact process to corresponding questions about 1—dependent site percolation with p close to 1, problems which usually can be solved using the contour argument of Section 5a. The first thing to do is to define the percolation process and describe its relationship to the contact process. Let V be the graph with vertices { (m,n} : m+n is even }, and oriented bonds connecting each (m,n) to (m+1,n+1) and to (m−1,n+1). Stealing a term from the physics literature, we call V the "renormalized lattice." To explain the name (and the idea behind the construction), the reader should imagine V mapped into the upper half plane $\mathbb{R} \times [0,\infty)$ by $\varphi(x,y) = (aLx,Ly)$, where a is a special constant and L is a large number to be chosen below.

We say that a site $(m,n) \in V$ is open if a certain "good event" happens in the graphical representation near $z_{m,n} = \varphi(m,n)$. We will do this in such a way that:

(i) the random variables $\eta(v)$, $v \in V$, which indicate whether the sites are open or not, are 1 dependent, that is, if we let $\|x{-}y\|_V = |x{-}y|/2$, then $\eta(x)$ and $\eta(y)$ are independent when $\|x{-}y\|_V > 1$;

(ii) if L is large the probability $\eta(x)$ is open is close to 1; and

(iii) if percolation occurs starting from 0 on the renormalized lattice, there is an infinite path starting near (0,0) in the graphical representation for the contact process.

As we go through the proof, the reader should notice that all we use is that $\alpha(\lambda) > 0$. Let A be the parallelogram with vertices $(-3\epsilon K,0)$, $(3\epsilon K,0)$, $((\alpha-3\epsilon)K,K)$, and $((\alpha+3\epsilon)K,K)$. From (2) in Section 4a, we know that $r_K/K \to \alpha$ as $K \to \infty$, so if $\delta > 0$ and $K \geq K_0(\delta,\epsilon)$, then $P(\ r_K \in ((\alpha-\epsilon)K,(\alpha+\epsilon)K)\ ) \geq 1-\delta$. When r_K is in $((\alpha-\epsilon)K,(\alpha+\epsilon)K)$ we know that there is a path from $(-\infty,0]\times\{0\}$ to $((\alpha-\epsilon)K,(\alpha+\epsilon)K)\times\{K\}$. Our next task is to show that if K is large, then with high probability any such path stays in A.

To prove that the path doesn't hit the right side, we observe that (2) in Section 4b implies

$$P(\ r_t > (\alpha+\epsilon)t\ ) \leq C_1\exp(-\gamma_1 t).$$

To turn this into the desired estimate, observe that comparing with a process with no deaths, in which the right edge R_t has a Poisson distribution with mean λt, gives

$$P(\ r_t > 2\epsilon K \text{ for some } t \leq \epsilon K/\lambda\ )$$
$$\leq P(\ R(\epsilon K/\lambda) > 2\epsilon K\ ) \leq C_2\exp(-\gamma_2 K),$$

and

$$P(\ r_t - r_n \geq \epsilon K \text{ for some } t \in [n,n+1]\ )$$
$$\leq P(\ R_1 \geq \epsilon K\ ) \leq C_3\exp(-\gamma_3 K).$$

Combining the last three estimates gives

$$P(\ r_t > \alpha t + 2\epsilon K \text{ for some } t \leq K)$$
$$\leq C_2\exp(-\gamma_2 K) + \sum_{n=\epsilon K/\lambda}^{K} C_1\exp(-\gamma_1 n) + KC_3\exp(-\gamma_3 K),$$

the three terms being estimates for (a) $r_t > 2\epsilon K$ at some $t \leq \epsilon K/\lambda$, (b) $r_t > \alpha t + \epsilon K$ at some integer time in the indicated range, and (c) $r_t > \alpha t + 2\epsilon K$ at some time t but $\leq \alpha[t] + \epsilon K$ at the previous integer time [t].

To prove that the path does not exit the left side, we observe that if there is a path that does this and ends in $((\alpha-\epsilon)K,(\alpha+\epsilon)K)\times\{K\}$, then the left edge of the dual starting from $[(\alpha-\epsilon)K,\infty)$ occupied must hit the left side of A, but looking in a mirror, and using the last calculation shows this has small probability. [This should explain why we have rectangles that go $3\epsilon K$ in each direction, and we estimated $P(r_t > \alpha t + 2\epsilon K$ for some $t \leq K$) above.]

The argument above shows that (ii) holds. The next step is to fit the renormalized bonds together so that (i) and (iii) do. The first step is to draw a picture of what we want (see Figure 11.1), and then we do a little arithmetic on scratch paper

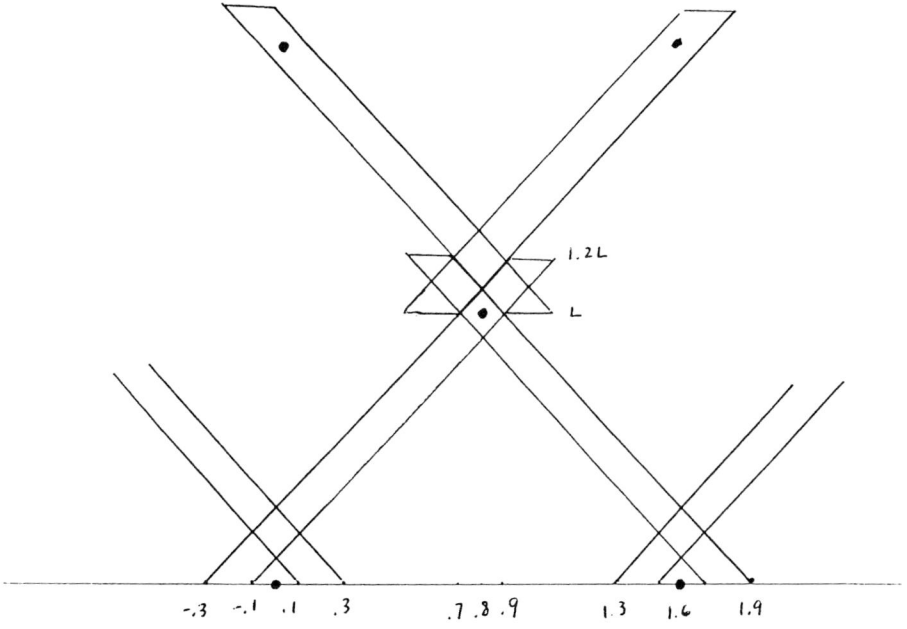

Figure 11.1

to figure out how to get this. A little trial and error leads to the following definitions.
Let $K = 1.2L$, and for $(m,n) \in V$ let

$$z_{m,n} = (\,(\alpha L - 3\epsilon K)m,\ Ln\,)$$
$$A_{m,n} = (\,z_{m,n} - (6\epsilon K,0)\,) + A$$
$$B_{m,n} = (\,z_{m,n} + (6\epsilon K,0)\,) - A$$
$$I_{m,n} = z_{m,n} + (\,(-9\epsilon K, 9\epsilon K) \times \{0\}\,)$$

where $c + A = \{\,c+x : x \in A\,\}$ and $c - A = \{\,c-x : x \in A\,\}$. In Figure 11.1 we have
drawn what happens when $\epsilon = \alpha/36$, i.e. $3\epsilon K = .1\alpha L$. This choice of ϵ is good enough
for the argument in this section, but for the second result in the next section, we will
need to choose ϵ small.

$A_{m,n}$ and $B_{m,n}$ are the "renormalized bonds" that appear in the title of the
section, and are thought of as being open if they contain a path of the desired type.
However, for technical reasons to be explained in a minute, we do not map the contact
process into bond percolation, but instead declare the site (m,n) to be open if there are
paths of the desired type in $A_{m,n}$ and $B_{m,n}$. To check (i), we observe that if the
parallelograms associated with two sites do not intersect, then the states of the sites
are independent, so the state of (m,n) is (positively) correlated with that of $(m-1,n-1)$,
$(m+1,n-1)$, $(m-2,n)$, $(m+2,n)$, $(m-1,m+1)$, and $(m+1,n+1)$, but independent of the
others. To check (iii), we look at a closeup of the region near $z_{1,1}$ to see (consult
Figure 11.2) that if $(0,0)$ and $(1,1)$ are open, then there is a path from $I_{0,0}$ to $I_{2,2}$ and
to $I_{0,2}$. The reader should note that to connect the paths in $A_{0,0}$ and $A_{1,1}$ we need the
path in $B_{1,1}$. This is why the renormalized system is chosen to be site rather than
bond percolation.

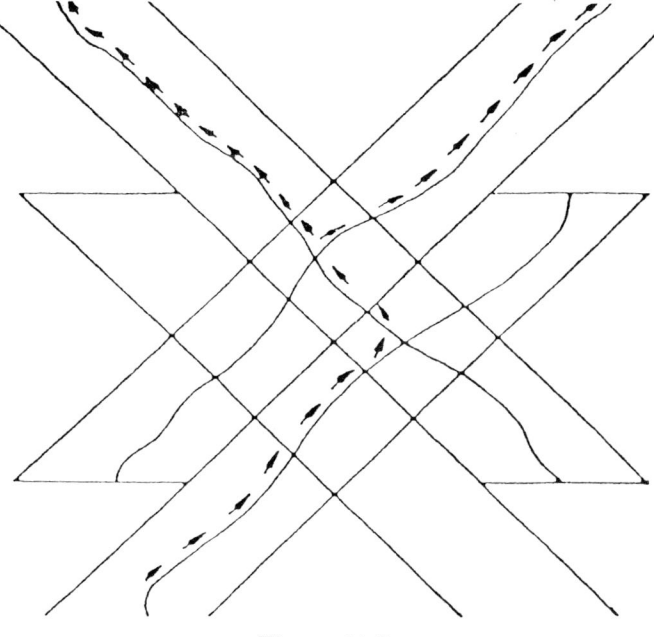

Figure 11.2

Having waded through the details of the construction, we can now start to reap the benefits. The easiest (and historically the first) consequence is

(1) Theorem. $\alpha(\lambda_c) = 0$.

Proof: Suppose $\alpha(\lambda_c) > 0$. Then since all the construction uses is the positivity of α (you were asked to check this as we went along), we can pick an L so that $P(\ \eta(z) = 1\ ) > 1 - 3^{-36}$, and hence by a result in Section 5a, there is positive probability of percolation in the η system. However, with L fixed, we are dealing with a bounded set, and it is easy to see that there is a $\lambda < \lambda_c$ so that $P(\ \eta(z) = 1\ ) > 1 - 3^{-36}$. But this implies there is positive probability of survival for this value of λ, contradicting the definition of λ_c.

Remark. The last proof may remind the reader of the proof in Section 6b that sponge crossing probabilities for site percolation have $\rho_{L,L}(p_c) \in [\epsilon_0, 1-\epsilon_0]$ (and hence there is no percolation at the critical value). This is no accident. Griffeath invented the construction above with the hope of proving that the conclusion in parentheses above held for the contact process. Unfortunately, it seems that one is only able to prove (1).

The idea of using the construction to study the supercritical contact process (something we will do in the next section) is due to Durrett and Griffeath (1983), but is inspired by the proofs at the end of Section 6a. Paths in parallelograms are our analogues of sponge crossings. The construction above is a distant relative of the original one, though. This is the sixth time the argument above has appeared in print, and the version above incorporates improvements contributed by Larry Gray, Tom Liggett, and Roberto Schonmann.

11b Exponential Estimates

In this section, we use the renormalized bond construction to prove that in the one dimensional contact process everything happens exponentially fast when $\lambda > \lambda_c$. Specifically, if we let $\tau^A = \inf\{\, t : \xi_t^A = \phi \,\}$ then we will show there are constants $C, \gamma \in (0,\infty)$ (which depend on λ but not on A) so that

(1)
$$P(\, \tau^A < \infty \,) \leq Ce^{-\gamma|A|},$$

(2)
$$P(\, t \leq \tau^A < \infty \,) \leq Ce^{-\gamma t}.$$

A reason for interest in the second inequality is that if ξ_t^1 denotes the system starting from $\xi_0^1 = \mathbb{Z}$, then duality implies

$$P(\, 0 \in \xi_t^1 \,) - P(\, 0 \in \xi_\infty^1 \,) = P(\, t \leq \tau^0 < \infty \,).$$

So (2) implies that ξ_t^1 converges to equilibrium exponentially fast. A second reason for interest in (2) is that it allows us to bound the correlations in the stationary distribution.

(3) Theorem. Let $\rho_t = P(\, \xi_t^1(x) = 1 \,)$. There are constants $C, \gamma \in (0,\infty)$ (which depend on λ but not on t or k) so that if $|x_1 - x_i| \geq m$, then for $1 < i \leq k$

$$\left| E \prod_{i=1}^{k} (\xi_t^1(x_i) - \rho_t) \right| \leq kCe^{-\gamma m}.$$

The last bound allows us to use the standard fourth moment proof of the strong law of large numbers to prove

(4) Theorem. As $t \to \infty$, $|\xi_t^0|/t \to 2\alpha\rho \, 1_{\Omega_\infty}$ a.s.

Having seen a strong law for $|\xi_t^0|$, the reader should wonder about the central limit theorem. Galves and Presutti (1987) have shown that if $\lambda > \lambda_c$ and $r_t = \sup \xi_t^{(-\infty,0]}$ then

$$(r_t - \alpha t)/t^{1/2} \Rightarrow \sigma_0 \, \chi$$

where χ is a standard normal and $\sigma_0 \in (0,\infty)$. From their result, it should follow that conditional on Ω_∞

$$(|\xi_t^0| - 2\alpha\rho t)/t^{1/2} \Rightarrow \sigma_1 \, \chi.$$

We leave the details as a publishable exercise for an interested reader. To get started , observe that: (i) a result of Newman and Wright (1981) implies

$$(|\xi_t^1 \cap [-\alpha t, \alpha t]| - 2\alpha\rho t)/t^{1/2} \Rightarrow \sigma_2 \, \chi$$

(for more details see Durrett (1984a), p.1034–1035), and (ii) recall the coupling result (5) from Section 4a. It is easy to see that $\sigma_1^2 = \sigma_2^2 + 2\rho^2\sigma_0^2$. The hard part is to show that the edge and density fluctuations are independent.

The rest of the section is devoted to the proofs of $(1) - (4)$. As advertised in the last section, the first two results will be proved by reducing them to corresponding problems about 1–dependent site percolation with p close to 1, which are then solved by using the contour argument of Section 5a.

Proof of (1): We begin with the special case A = {0, 1, ... n}. In the renormalized bond construction let $\epsilon = \alpha/36$, and pick L so that $P(\eta(z) = 1) > 1 - 3^{-72}$. (3) in Section 5a implies that if we start with $\{0, -2, ... -2N\}$ occupied, the probability of percolation in η system is at least $1 - C \cdot 9^{-N}$. The construction guarantees that when this happens there is a path in the contact process starting in

$$\bigcup_{m=0}^{-2N} I_{m,0} \subset [-1.6\alpha LN - .3\alpha L, .3\alpha L],$$

so we have shown

$$P(\tau^{\{0,...,(1.6N+.6)\alpha L\}} < \infty) \le C \cdot 9^{-N},$$

which implies the desired result for A = {0, 1, ... n}.

To prove the result for a general A, one shows that if $|A| = n$ then

(5) $$P(\tau^A < \infty) \le P(\tau^{\{1,...,n\}} < \infty).$$

This inequality is due to Tom Liggett. It says that if we spread the particles out more, then the chance of survival increases, and is proved by a construction that makes the intuition precise. One defines $\xi_t^{\{1,...,n\}}$ and ξ_t^A on the same space so that there is always a map φ_t from $\xi_t^{\{1,...,n\}}$ into ξ_t^A with $|\varphi_t(x) - \varphi_t(y)| \ge |x-y|$ for all t, and $x,y \in \xi_t^{\{1,...,n\}}$. This shows $|\xi_t^A| \ge |\xi_t^{\{1,...,n\}}|$, and proves (5). The construction is left as an exercise for the reader — there is only one reasonable thing to try, and it works. For a solution, see page 10 of Durrett and Griffeath (1983).

Proof of (2): To prove this, we begin by proving a large deviations result for the right edge which is of interest in its own right:

(6) Theorem. If $\lambda > \lambda_c$ and a $< \alpha(\lambda)$, there are constants $C, \gamma \in (0, \infty)$ (which depend on λ and a) so that

$$P(\, r_t < at \,) \leq Ce^{-\gamma t}.$$

Proof: We begin by considering the analogous problem for 1–dependent site percolation. Let $H = \{ \, y \in V :$ for some $(m,0)$ in V with $m \leq 0$ there is a path of open sites from $(m,0)$ to $y \, \}$, and let $s_n = \sup\{ \, m : (m,n) \in H \, \}$. The first step is to prove

(7) Lemma. If $\theta < 1$ and $p > 1 - 3^{-72/(1-\theta)}$, then

$$P(\, s_n < n\theta \,) \leq 3^{-n}.$$

The proof of (7) involves a contour argument. Let $W = \cup_{y \in H} y + D$ where $D = \{ \, (a,b) : |a| + |b| \leq 1 \, \}$, and let $\Gamma =$ the boundary of the unbounded component of $W^c \cap (\, \mathbb{R} \times (0,n) \,)$. Γ runs from $(1,0)$ to (s_n+1,n). (See Figure 11.3.) Orient Γ so that $(1,0)$ is at the beginning, and let N_1, N_2, N_3, and N_4 be the number of segments of Γ (of length $\sqrt{2}$) that point NW, SW, NE, and SE. Now $N_3 + N_4 - N_1 - N_2 = s_n$, so if Γ has length n+k and $s_n < \theta n$ we have

$$2(N_1 + N_2) = N_1 + N_2 + N_3 + N_4 - s_n \geq n+k - \theta n,$$

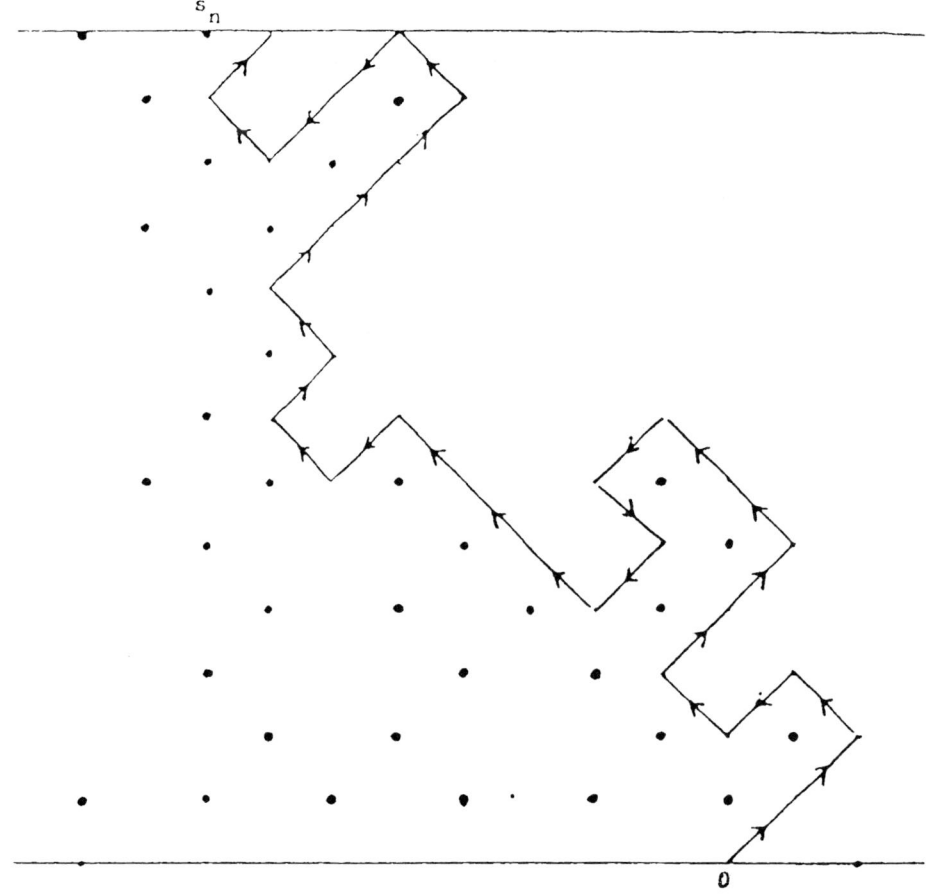

Figure 11.3

which implies $N_1 + N_2 \geq (1-\theta)(n+k)/2$. Since segments of type 1 and 2 move us to the left, it follows as in Section 5a that there are at least $(N_1 + N_2)/2$ sites that are closed, and there is a subset of these sites of size $\geq (N_1 + N_2)/18$ which are independent. Since there are at most $2 \cdot 3^{n+k-1}$ contours with $n+k$ edges (the first segment is either $(1,0) \to (2,1)$ or $(1,0) \to (0,1)$), and $(1-p) < 3^{-72/(1-\theta)}$, we arrive at the estimate

$$P(\ s_n < \theta n\ ) \le \sum_{k=0}^{\infty} 2 \cdot 3^{n+k-1}(1-p)^{(1-\theta)(n+k)/36}$$

$$\le (2/3) \sum_{j=n}^{\infty} 3^{-j} = 3^{-n},$$

proving (7).

To prove (6) now, choose $\delta < \alpha - a$, and pick $\theta < 1$ so that $\theta(\alpha - \delta) > a$. Take $\epsilon = \delta/3.6$ in the renormalized bond construction, so that $\alpha L - 3\epsilon K = (\alpha - \delta)L$ (recall $K = 1.2L$), and the renormalized lattice is $z_{m,n} = ((\alpha - \delta)Lm, Ln)$. Pick L so that in the η system $P(\ \eta(z) = 1\ ) > 1 - 3^{-72/(1-\theta)}$. If s_n is defined for the η system as above, and r_t is the right edge of the contact process, then $r_t \ge (\alpha - \delta)L(s_n - 1) - 3\delta L$ for all $t \in [(n-1)L, nL]$. (In the worst case there is a path in $A_{s(n)-1,n-1}$. Since $9\epsilon K = 3\delta L$, the right-hand side is the leftmost point in this parallelogram.) Since $\theta(\alpha - \delta) > a$, the desired result follows from the last inequality and (7).

Remark. While on the subject, it seems worthwhile to notice that (6) can be improved to

(6') $$\lim_{t \to \infty} \frac{1}{t} \log P(\ r_t < at\ ) = -\gamma(a) < 0.$$

Using the notation of (1) from Section 4a, $r_s + r_{s,s+t} \ge r_{s+t}$, where $r_{s,s+t}$ is independent of r_s and has the same distribution as r_t, so

$$P(\ r_{s+t} < a(s+t)\ ) \ge P(\ r_s < as, r_{s,s+t} < at\ ) = P(\ r_s < as\ )\ P(\ r_t < at\ ).$$

Taking logs, we see that $f(t) = \log P(\ r_t < at\ )$ is superadditive, so consulting the proof of (1) in Section 4b, we see $f(t)/t \to \sup f(s)/s$. So (6') holds and, in addition, we have

$$P(r_t < at) \le e^{-\gamma(a)t}.$$

The last trick does not work for $P(r_t > bt)$ with $b > a$, but a related one does, at least for $\lambda > \lambda_c$. Let $r_t^0 = \sup \xi_t^0$. Then

$$P(r_{s+t}^0 > b(s+t)) \ge P(r_s^0 > bs)P(r_t^0 > bt),$$

so $(1/t)\log P(r_t^0 > bt)$ has a limit, and we get the result for r_t by noting

$$P(r_t > bt) \ge P(r_t^0 > bt) \ge P(r_t > bt, \Omega_\infty) \ge P(r_t > bt) P(\Omega_\infty),$$

the last inequality following from Harris' inequality from Chapter 6. (As stated there his result does not apply to the graphical representation. However, in Section 5c, we showed that our graphical representations are limits of percolation processes so a simple limiting argument, left to the reader, shows its application is valid here.)

Returning to our main topic, our next step in proving (2) is to show

(8) $$P(t \le \tau^0 < \infty) \le Ce^{-\gamma t}.$$

Proof: If we let $\ell_t = \inf \xi_t^{[0,\infty)}$, then

$$P(t \le \tau^0 < \infty) \le P(r_s < 0 \text{ or } \ell_s > 0 \text{ for some } s \ge t).$$

Using (6) gives that $P(\, r_s < 0\,) \leq Ce^{-\gamma s}$, and integrating gives

$$E|\{\, s \geq t : r_s < 0\,\}| \leq (C/\gamma)e^{-\gamma t}.$$

To get the desired result from this, observe that

$$E|\{\, s \geq t : r_s < 0\,\}| \geq (1/\lambda)P(\, r_s < 0 \text{ for some } s \geq t),$$

since r_s will remain < 0 at least until the next arrow from -1 to 0.

(8) is (2) with $A = \{0\}$. To get the general result, we can use a restart argument like the one in the proof of (6) in Section 4b. As long as $\xi_t^A \neq \phi$, we keep looking for a particle that starts a process that lives forever. In order for $t \leq \tau^A < \infty$, we must not find one before time t. A little computation shows that (2) holds with a γ that is $\rho = P(\, \tau^0 = \infty\,)$ times the γ in (8). We spare the reader the details, which are given in Section 12 of Durrett (1984a). We will only use the result for $A = \{0\}$ below.

The next item on our agenda is to prove (3). We will prove the result for the variables

$$\zeta_t(x) = 1_{(\, \xi_t^x \neq \phi\,)} - \rho_t,$$

which have the same joint distribution. The argument splits into two cases: $t \leq m/2\lambda$ and $t > m/2\lambda$. To handle the first case, let

$$G(x) = \{\, \xi_s^x \subset [-m+x, m+x] \text{ for all } s \leq m/2\lambda\,\}, \qquad G = \bigcap_{i=1}^{k} G(x_i),$$

and write

$$|E(\prod_{i=1}^{k} \zeta_t(x_i))| \le P(G^c) + |E(\prod_{i=1}^{k} \zeta_t(x_i), G)|.$$

(Notice the product is ≤ 1.) A by now familiar, comparison with a rate λ Poisson process yields $P(G^c) \le kCe^{-\gamma m/2\lambda}$. Conditional on G, $\zeta_t(x_1)$ and $(\zeta_t(x_2),...\zeta_t(x_k))$ are independent when $t \le m/2\lambda$, so the second term on the right is majorized by

$$|E(\prod_{i=1}^{k} \zeta_t(x_i)|G)| = |E(\zeta_t(x_1)|G)||E(\prod_{i=2}^{k} \zeta_t(x_i)|G)|$$

$$\le |E(\zeta_t(x_1)|G)|$$

$$= |E(\zeta_t(x_1),G)|/P(G)$$

$$= |E(\zeta_t(x_1),G^c)|/P(G)$$

$$\le P(G^c)/P(G),$$

where in the last equality we have used $E\zeta_t(x_1) = 0$. Multiplying the last string of inequalities by $P(G)$, and combining with previous results, we have

$$|E(\prod_{i=1}^{k} \zeta_t(x_i))| \le 2P(G^c),$$

completing the proof for $t \le m/2\lambda$. To prove the result for $t \ge m/2\lambda$, let

$$\zeta_t'(x) = 1_{(\xi_{m/2\lambda}^x \ne \phi)} -\rho_t$$

$$\zeta_t''(x) = 1_{(\xi_{m/2\lambda}^x \ne \phi)} -\rho_{m/2\lambda}$$

By (8),

$$P(\zeta_t'(x) \ne \zeta_t(x)) \le P(m/2\lambda \le \tau^0 < \infty) \le Ce^{-\gamma m/2\lambda},$$

and hence,

$$E| \prod_{i=1}^{k} \zeta_t(x_i) - \prod_{i=1}^{k} \zeta_t'(x_i)| \le 2kCe^{-\gamma m/2\lambda}.$$

Now if we have numbers $-1 \le a_i, b_i \le 1$, an easy induction argument shows

$$| \prod_{i=1}^{k} a_i - \prod_{i=1}^{k} b_i| \le \sum_{i=1}^{k} |a_i - b_i|,$$

and so

$$E| \prod_{i=1}^{k} \zeta_t'(x_i) - \prod_{i=1}^{k} \zeta_t''(x_i)| \le k \, P(\, m/2\lambda \le \tau^0 < \infty \,).$$

Finally the argument for $t \le m/2\lambda$ gives

$$E| \prod_{i=1}^{k} \zeta_t''(x_i)| \le Ce^{-\gamma m/2\lambda},$$

and combining the last three estimates gives the result for $t > m/2\lambda$.

Last, but not least, we want to prove (4). In the proof of (6) in Section 4b, we observed that

$$\xi_t^0 = [\ell_t^0, r_t^0] \cap \xi_t^1 \approx [-\alpha t, \alpha t] \cap \xi_t^1,$$

so we will begin by considering the limiting behavior of the last quantity on the right. Let $\rho_t = P(\, x \in \xi_t^1 \,)$, let $\zeta_t(x) = \xi_t(x) - \rho_t$, and let S_t be the sum of $\zeta_t(x)$ for $-\alpha t \le x \le \alpha t$. Now

$$E(S_t)^4 \le 4! \, \Sigma \, |E \, \zeta_t(w)\zeta_t(x)\zeta_t(y)\zeta_t(z)|$$

where the sum is over $-\alpha t \leq w \leq x \leq y \leq z \leq \alpha t$. The number of terms with $\max(|w-x|,|y-z|)$
$= 2m$ is at most $(2\alpha t+1)^2 2(2m+1)$ — the number of ways to pick w and z is at most
$(2\alpha t+1)^2$, and having picked w and z, there are at most $2(2m+1)$ ways to pick x and
y. (Notice the max $= 2m$, and x and y lie between w and z.) If the max above is
attained by $|w-x|$, we apply (3) with $x_1 = w$, otherwise we take $x_1 = z$. It follows
that

$$E(S_t)^4 \leq (24 \sum_{m=0}^{\infty} (4m+2)Ce^{-\gamma m})(2\alpha t+1)^2 \leq C't^2$$

(the last inequality holding for $t \geq 1$).

Letting $t=1,2,\dots$ and summing, shows that $S_n/n \to 0$ almost surely when $n \to \infty$
through the integers. Extending the last result to conclude that $S_t/t \to 0$ as $t \to \infty$ is
neither difficult nor interesting, and is left to the reader. (See Durrett and Griffeath
(1983), p.14 for details.) Having skipped the tedious part of the proof, the rest is easy.
On Ω_∞, we have $\ell_t^0/t \to -\alpha$ and $r_t^0/t \to \alpha$ a.s., and there is at most one particle per site,
so the difference between $|\xi_t^0| = |\xi_t^Z \cap [\ell_t^0, r_t^0]|$ and $|\xi_t^Z \cap [-\alpha t, \alpha t]|$ is o(t) on Ω_∞.

11c A General Shape Theorem

In this section, we state and prove a general "shape theorem" due to Durrett and Griffeath (1982) (a reference that we will hereafter abbreviate as DG(82)). This theorem generalizes the result proved for Richardson's model in Chapter 1, and will be applied in the next section to the biased voter model and contact process. We begin with a number of definitions. Let $S = \{$ all subsets of $Z^d \}$. A family $\{ \xi_t^A : A \subset Z^d \}$ of S valued Markov processes is said to be a "growth model" if ϕ is an absorbing state and the family is:

(i) *translation invariant* : the translated process $x + \xi_t^A$ is a copy of ξ_t^{x+A} (here $x + B = \{ x+y : y \in B \}$);

(ii) *attractive* : if $A \subset B$ then ξ_t^A and ξ_t^B can be constructed on the same space in such a way that $\xi_t^A \subset \xi_t^B$ for all t; and

(iii) *finite range* : there is an $L < \infty$ so that if $A \cap \{ x : |x| \leq L \} = \phi$ then $P(0 \in \xi_t^A)$ $= o(t)$ as $t \to 0$.

Richardson's model, the biased voter model, and the contact process clearly satisfy all these conditions. Since we will only be interested in these examples, we will suppose for the proof that (iii) holds with $L = 1$, and that the processes under consideration can be constructed from a graphical representation, although neither of these assumptions is necessary for the theorem we are about to state. Let

$$\tau^A = \inf\{ t : \xi_t^A = \phi \}, \qquad t^A(x) = \inf\{ t : x \in \xi_t^A \}.$$

Our attention will be focused on the growth process ξ_t^0 that starts from $\{0\}$, so we will

let $\tau = \tau^0$ and $t(x) = t^0(x)$. We say that ξ_t is supercritical if $P(\tau = \infty) > 0$. In

formulating the shape theorem it will be convenient (as it was in Chapters 1,8, and 9)

to identify each $x \in \mathbb{Z}^d$ with the cube with side one having center at x. With this in

mind, we define the set of sites hit by time t to be

$$H_t = \{\, y \in \mathbb{R}^d : \text{there is an } x \in \mathbb{Z}^d \text{ with } \|x{-}y\| \leq 1/2 \text{ and } t(x) \leq t\,\}.$$

(Here and throughout this section we will use the L^∞ norm on $\mathbb{R}^d$.) Now, as we

mentioned in Section 5c, any attractive system has a stationary distribution ν which is

the limit of ξ_t^1, the process starting from $\xi_0^1 = \mathbb{Z}^d$. To help us prove that

$$\xi_t^0 \Rightarrow \delta_\phi \, P(\tau^0 < \infty) + \nu \, P(\tau^0 = \infty),$$

we introduce the coupled region

$$K_t = \{\, y \in \mathbb{R}^d : \text{there is an } x \in \mathbb{Z}^d \text{ with } \| x{-}y \| \leq 1/2 \text{ and } \xi_t^0(x) = \xi_t^1(x)\}.$$

Here and below, we use the coordinate notation $\xi_t^A(x) = 1$ if $x \in \xi_t^A$, $= 0$ otherwise.

Very loosely, ξ_t^0 is "in equilibrium" on K_t. To see how loose the last statement is,

observe that $K_t \supset \{\, x : \xi_t^1(x) = 0 \,\}$ and $\xi_t^1 \neq \nu$!

The next theorem is the main result of this section. It gives sufficient conditions

for ξ_t^0 to have an asymptotic shape. The conditions are that three sorts of probabilities

should decay exponentially fast in time.

(1) **Theorem.** Let ξ_t^0 be a supercritical growth model. Suppose there are constants $C, c, \gamma \in (0, \infty)$ so that

(a)
$$P(\, t \leq \tau < \infty \,) \leq Ce^{-\gamma t}$$

(b)
$$P(\, t(x) > t, \tau = \infty \,) \leq Ce^{-\gamma t} \quad \text{for } \|x\| < ct.$$

Then there is a convex set A so that on $\{\, \tau = \infty \,\}$ we have for any $\epsilon > 0$

(∗)
$$(1-\epsilon)tA \subset H_t \subset (1+\epsilon)tA$$

for all t sufficiently large. If, in addition,

(c)
$$P(\, x \notin K_t, \tau = \infty \,) \leq Ce^{-\gamma t} \quad \text{for } \|x\| < ct,$$

then on $\{\, \tau = \infty \,\}$ we have for any $\epsilon > 0$

(∗∗)
$$(1-\epsilon)tA \subset (H_t \cap K_t) \subset (1+\epsilon)tA$$

for all t sufficiently large.

At this point several remarks are in order. First we must intersect K_t with H_t, because $K_t \supset \{\, x : \xi_t^1(x) = 0 \,\}$. Second, we cannot have coupling on a much larger set because $\xi_t^0(x) = 0$ on H_t^c. Third, it is enlightening to compare the conclusions here with those for the one dimensional contact process. There

$$H_t = [\inf_{s \le t} \ell_s^0, \sup_{s \le t} r_t^0] \quad \text{and} \quad \xi_t^0 = \xi_t^1 \cap [\ell_t^0, r_t^0], \quad \text{so} \quad K_t \supset [\ell_t^0, r_t^0],$$

and it follows from results in Section 4a that

$$H_t/t \quad \text{and} \quad (H_t \cap K_t)/t \to [-\alpha(\lambda), \alpha(\lambda)].$$

Finally, as in the case of the one dimensional contact process, the result above implies

$$\xi_t^A \to \delta_\phi \, P(\, \tau^A < \infty \,) + \nu \, P(\, \tau^A = \infty \,)$$

so all stationary distributions are of the form $\theta \, \delta_\phi + (1-\theta) \, \nu$. As the reader may have noticed, our last remark is STUPID because the result we mentioned follows from hypothesis (c) in (1), but it did give us a chance to mention the complete convergence theorem. The strong law

$$|\xi_t|/t^d \to \rho |A| 1_{\{\, \tau = \infty \,\}} \quad \text{a.s.}$$

(where $\rho = P(\, x \in \xi_t^\nu \,)$ and $|A|$ is the volume of A) is an example of a useful result that follows from the theorem but is not an immediate consequence of its hypotheses.

Remark: Before starting on the proof of (1), we would like to illustrate the result by a computer simulation. In Figure 11.4, we have simulated the two dimensional contact process with $\lambda = 1$ and $A = \{-1,0,1\}^2$. (Recall, Holley and Liggett (1978) have shown $\lambda_c \le 2/d$, and computer simulations suggest $\lambda_c \approx .41$. See the discussion at the end of Section 4a.) 1's (resp. 0's) indicate places where $\xi_t^A(x) = \xi_t^1(x) = 1$ (resp. 0). x's mark points in $H_t - K_t$. Finally, .'s mark points in $\xi_t^1 \cap H_t^c$.

Figure 11.4   Time = 50   Density = .7212

Proof of (1): We will describe the main ingredients that go into the proof, leaving the reader to find the rest of the gory details in DG(82). We hope that the outline and intuition provided below will help the reader to digest that paper. To facilitate reading DG(82), we have kept the notation used there with some minor changes, the most notable being that we use ξ here instead of η. At two places below we will depart from the argument in DG(82). In one case, we make the argument simpler. In the other, we make it correct.

As we mentioned earlier, we will carry out the proof only for processes that can be constructed on a graphical representation. The reason for this change is that we can define the process starting from A occupied at time s to be

$$\xi_t^{(A,s)} = \{ \, y : \text{there is a path from (x,s) to (y,t) for some } x \in A \, \}$$

for $t \geq s$ (and $= \phi$ for $t < s$), and it will follow that if $\xi_s^B \supset A$ then $\xi_t^B \supset \xi_t^{(A,s)}$. In the attractive case, one can always define the last two processes on the same space in such a way that the last inequality holds, but when there is a graphical representation we can define all the processes we will ever need at one time.

We will only prove (∗) below. The proof of (∗∗) is very similar. To prove (∗), we will show

I. (almost) subadditivity

$$t(x+y) \leq t(x) + s(y) + v(x,y)$$

where s(y) is an appropriately chosen copy of t(y) that is independent of t(x), and v(x,y) is an error term with

$$\overline{E}(\ v(x,y)^2\ ) = O(\ \|x+y\|\ ).$$

Here $\overline{E}(\ ) = E(\ |\ \tau = \infty\ )$, and $\overline{P}(\ ) = P(\ |\ \tau = \infty\ )$.

II. For some $C, c, \gamma \in (0,\infty)$

$$\overline{P}(\ B_{x,ct} \nsubseteq H_{t(x)+\ell^2+t}\ \text{for some}\ t \ge 0\ ) \le Ce^{-\gamma\ell},$$

where $B_{x,r} = \{\ y : \|\ x-y\ \| \le r\ \}$ is the box of radius r centered at x. (Recall $\|\ \|$ is the L^∞ norm.)

If we let $x = 0$ in II, then we see $H_t \supset B_{0,ct/2}$ for all t sufficiently large, a conclusion that should be familiar from Chapters 1, 3, and 8. It is not hard to show that II implies there is a $C \in (0,\infty)$ so that

(2) $$\overline{E}(\ t^2(kx)\ ) \le Ck^2\|x\|^2 + O_x(k)$$

where $O_x(k)$ denotes a term that is $\le C(x)\ k$ and, as indicated, the constant depends on x. The original result of Richardson (1973), when supplemented by Kesten's contribution to the discussion of Kingman (1973), shows that I and (2) imply radial limits exist.

(3) As $n \to \infty$, $t(nx)/n \to \mu(x)$ $\overline{P}$ a.s.

Here the passage times have been extended to $x \in \mathbb{R}^2$ in the usual way ($t(x) = \inf\ \{\ t : x \in H_t\ \}$). We leave the proof of (3) as an exercise for the reader. Imitate the proof in

Chapter 1 or look up the answer in Richardson (1973). For an almost subadditive ergodic theorem, see Derriennic (1983).

We observed in the Denouement of Chapter 1 that II and (3) imply (∗), so it remains to demonstrate I and II. We will prove II first and then I. The proofs involve a lot of tedious computations. To keep unpleasant details in the background as much as possible, γ,c, and C will denote constants whose vales are unimportant, and in general will change from line to line.

(4) Assuming (a) and (b), II holds.

Proof: If $\| x \| \leq c\ell^2$, then the verification is easy:

$$\overline{P}(B_{x,ct} \not\subseteq H_{t(x)+\ell^2+t} \text{ for some } t \geq 0) \leq \overline{P}(B_{x,ct} \not\subseteq H_{\ell^2+t} \text{ for some } t \geq 0)$$
$$\leq \overline{P}(B_{0,cs} \not\subseteq H_s \text{ for some } s \geq \ell^2).$$

Using (b) now gives that the last quantity is $\leq Ce^{-\gamma\ell^2}$.

With the trivial case out of the way, we can assume for the rest of the proof that $\| x \| > c\ell^2$. Let T be the time at which $\overline{\xi}_t^0 = (\xi_t^0 \mid \tau = \infty)$ first hits $B_{x,\ell}$ (The inside box $B_{x,\ell/2}$ used in DG(82) is not necessary.) Suppose we can construct:

(5a) a random time σ so that $\overline{P}(\sigma \geq t(x)) \leq Ce^{-\gamma\ell}$; and

(5b) a copy $\tilde{\xi}_{\sigma+t}$ of $\overline{\xi}_t^y$, for some random $y \in B_{x,\ell}$, with $\tilde{\xi}_{\delta+t} \subseteq \overline{\xi}_{\sigma+t}^0$ $\overline{P}$ a.s.

Then the last two conclusions imply that with "overwhelming probability" (i.e., at least $1 - Ce^{-\gamma\ell}$), $\bar{\xi}_{\sigma+t}$ starts at $y \in B_{x,\ell}$ at time $\sigma \le t(x)$. The point y is within a distance ℓ of x, so given the time lag ℓ^2 it will, in view of assumption (b), cover $B_{y,\ell+ct} \supset B_{x,ct}$ at time $t(x) + \ell^2 + t$ with overwhelming probability.

To finish the proof of (4), it suffices to construct σ and check (5a) and (5b). The method is crude; our old friend the restart construction. Once the process enters $B_{x,\ell}$ we pick a point. If the process that it starts dies out, we wait until ξ_t^0 reenters $B_{x,\ell}$ and pick another point. After a geometrically distributed number of trials, we will find a process that lives forever. The random time at which this happens will be σ, and the lucky point will be y. To carry out this idea let $v_0 = 0$ and for $k \ge 1$ make the following definitions.

Let $u_k = \inf \{\, t \ge v_{k-1} : \xi_t^0 \cap B_{x,\ell} \ne \phi \text{ or } \xi_t^0 = \phi \,\}$.

If $\xi_{u(k)}^0 \cap B_{x,\ell} \ne \phi$ let x_k be a randomly chosen point in that set,

    otherwise (i.e., when $\xi_{u(k)}^0 = \phi$) let $x_k = 0$.

Let $\xi_t^k = \xi_t^{(x(k),u(k))}$, that is, the process starting from x_k occupied at time u_k.

Let $v_k = \sup\{\, t : \xi_t^k \ne \phi \,\}$, and $K = \inf\{\, k : v_k = \infty \,\}$.

Let $y = x_K$, and $\bar{\xi}_t = \xi_t^K$.

Notice the absence of the bar over ξ_t^0 in the above, and that even after $\xi_t^0 = \phi$ we go on looking for a process that lives forever. We do this so we can claim

(6a) the distribution of K under P is geometric (each restart is the last with probability $\rho = P(\, \tau = \infty \,)$); and

(6b) given $K = k$, the $v_j - v_{j-1}$, $1 \leq j \leq k-1$, are i.i.d. with a distribution that has an exponential tail by (a).

To see that (6a) and (6b) are correct, note that the selection of x_k is measurable with respect to the graphical representation at time $t \leq u_k$, and we wait to see if the chosen process dies out before we pick again. In DG(82) it is claimed that the last two conclusions hold under $\overline{P}$, but this is WRONG! The conditioning on $\{ \tau = \infty \}$ causes the success probability in (6a) to be $> \rho$, and does unpredictable things to the probability of $\{ t \leq v_1 < \infty \}$ and the dependence between the random variables.

Returning to our proof, we observe that the conclusion in (5b) holds by construction. To check (5a), we introduce the occupation times

$$\psi(t) = \int_0^t 1 (\xi_s^0 \cap B_{x,\ell} \neq \phi \text{ or } \xi_s^0 = \phi) \, ds,$$

and show that for some $C, \gamma, \epsilon \in (0,\infty)$

(7a) $P(\psi(\sigma) \geq m) \leq Ce^{-\gamma m}$,

(7b) $P(\psi(t(x)) \leq \epsilon \ell) \leq Ce^{-\gamma \ell}$,

from which (5a) follows immediately. (Set $m = \epsilon \ell$.) We use the occupation time $\psi(t)$ instead of ordinary time in (7), because of the following nightmare. The box $B_{x,\ell}$ is reached very quickly by a chain of events in which particles give birth very quickly and die soon after childbirth. If the particle that first lands in $B_{x,\ell}$ dies without having children, then we may have to wait an enormous amount of time for ξ_t^0 to reenter $B_{x,\ell}$.

but if we measure time with $\psi(t)$, the clock doesn't start until the reentry occurs (or until $\xi_t^0 \neq \phi$ and the meaningless part of the construction takes over).

Having explained the philosophy we turn now to the details. To prove (7a) we observe that

$$\psi(\sigma) = \sum_{j=1}^{K-1} v_j - v_{j-1},$$

and that if $\varphi(\theta) = E \exp(\theta v_1)$, then (6a) and (6b) imply

$$E \exp(\theta\psi(\sigma)) = \sum_{k=0}^{\infty} (1-\rho)^k \rho \, \varphi(\theta)^k = \frac{\rho}{1-(1-\rho)\varphi(\theta)} < \infty,$$

for θ small. So

$$P(\, \sigma \geq m \,) \leq e^{-\theta m} \, E \exp(\theta\sigma),$$

proving (7a).

To prove (7b), we begin by observing that if we let η_t be Richardson's model starting with everything in $B_{x,\ell}^c$ occupied, and run at a rate $\Lambda = $ the maximum birth rate in ξ_t (i.e. the birth rate at 0 when we start with 0 vacant and all other sites occupied), then the probability η_t reaches x by time $\epsilon\ell$ is

(8) $\leq |\partial B_{x,\ell}| \, (2d)^\ell \, P \begin{bmatrix} \ell \text{ mean } 1/\Lambda \text{ exponentials} \\ \text{sum up to less than } \epsilon\ell \end{bmatrix},$

which $\leq Ce^{-\gamma t}$ if ϵ is small. If we let $T(t) = \sup \{ s : \psi(s) \leq t \}$ be the inverse of $\psi(t)$, then an easy argument shows $\xi_{T(t)}^0 \subset \eta_t$. (When $B_{x,\ell}$ is entered for the first time, the point is on $\partial B_{x,\ell}$ and η_0 contains all these points. The second and subsequent times that ξ_t enters $B_{x,\ell}$ the discrepancy is worse, since η_t has grown in the interim.)

(7b) is the last piece in the proof of II, so we turn now to the proof of I. This time we will just give the proof "in words", and leave the reader to look up the details in DG(82), see pages 543–544.

(9) Assuming (a) and (b), I holds.

Proof: Fix $x,y \in Z^d$ and write $\ell = \|x\|^{1/2}$. First, using the restart construction from the proof of II, we find a time σ, and a one site process $\tilde{\xi}_{\sigma+t}$ that starts in $B_{x,r}$ and lives forever. From (5a) it follows that $P(\ \sigma \geq t(x)\ ) \leq Ce^{-\gamma\ell}$. Now there are two cases: $\sigma \leq t(x)$ and $\sigma > t(x)$. When $\sigma \leq t(x)$ (the good case), we begin a restart construction at time $t(x)$ to find a time $T \geq t(x)$, and a one site process ζ_{T+t} that lives forever and is imbedded in $\tilde{\xi}$ (the process constructed in the proof of II). When $\delta > t(x)$ (the bad but rare case), we look for a process ζ_{T+t} that lives forever and is embedded in ξ (the original process).

In either case we let $\{z\} = \zeta_T$ and define $s(y) = \inf\{t: z+y \in \zeta_{T+t}\ \}$. Clearly, $s(y)$ has the same distribution as $t(y)$ and is independent of $t(x)$. (This is why we wait until after $t(x)$ to start our restart construction.) Moreover, if we define

$$v(x,y) = (T{-}t(x)) + (t(x+y) - T - s(y))^{+} \equiv v_1 + v_2,$$

then

$$t(x+y) \leq t(x) + s(y) + v(x,y).$$

We now proceed to estimate v_1 and v_2 in order to show that $E(v^2(x,y)) = O(\ \|x\|\ )$. v_0 is a sum of a geometric number of i.i.d. random variables with the same distribution as

$(\tau \mid \tau < \infty)$, so (a) implies $E(v_0^2) \leq C$ independent of x. Thus, it suffices to handle v_1, which we break into two:

$$v_2 = v_1 \, 1_{(\sigma \leq t(x))} \text{ and } v_3 = v_1 \, 1_{(\sigma > t(x))}.$$

To bound $E(v_2^2)$, we first show that z is close to x with overwhelming probability. To do this, we first show $T-\sigma$ is small by observing we have shown that $\sigma-t(x)$ is small, and using II to show that soon after time σ, the process $\tilde{\xi}_{\sigma+t}$ will hit x. Comparing with Richardson's model shows that $\tilde{\xi}_T$ is not very big, and when the smoke clears we have

$$P(\parallel z-x \parallel > K(\ell+m^2) , \sigma \leq t(x)) \leq Ce^{-\gamma m}.$$

At time $T+s(y)$ the point $z+y$ is occupied, so using II at the point $z+y$ leads to

$$P(v_2 > K(\ell+m^2)) \leq Ce^{-\gamma m}.$$

To estimate v_3 is even easier. The bad case has probability $\leq Ce^{-\gamma \ell}$. Comparing with Richardson's model shows ξ_t expands at most linearly, so $\parallel z \parallel \leq K \parallel x \parallel$ with high probability, and using II again gives

$$E(v_3^2) \leq C\parallel x\parallel \, e^{-\gamma \ell} = o(\parallel x\parallel)$$

since $\ell = \parallel x\parallel^{1/2}$.

11d Checking the Hypotheses

We begin this section with an anecdote that will set the tone for the developments to follow. On "Four Way Street", a 1971 album of concert performances by Crosby, Stills, Nash, and Young, Neil Young starts to play "Don't let it bring you down," stumbles over the first few chords, and stops. Embarrassed, he says that the song "starts out real slow and then fizzles out altogether." The same statement could be applied to this section. We start to check the hypotheses of the shape theorem for the biased voter model. The first is no problem, but we fall a little short of proving the other two, and try to save face by arguing that we have done enough to prove the result. Having stumbled over the biased voter model, the section fizzles out altogether when we come to the contact process. We mumble a few words about the ideas behind the proof, give up, and change the subject to talk about what is known and what we would like to know. The last three paragraphs of this section are devoted to "applications" in an attempt to give the book a happy ending.

Having described our plan, we will now carry it out. In order to complete the proof of the shape theorem from Chapter 3, we would like to check that when $\lambda > 1$ the biased voter model satisfies (a) $-$ (c) of (1) in the last section. The first thing to show is:

(1) **Lemma.** If $\lambda > 1$ and $\tau = \inf\{t: \xi_t^0 \neq \phi\}$, then there are constants $C, \gamma \in (0, \infty)$ so that

$$P(\, t \leq \tau < \infty) \leq Ce^{-\gamma t}.$$

Proof: Since $|\xi_t^0|$ is a simple random walk run at rate $(\lambda+1)|\partial\xi_t^0|$ and $|\partial\xi_t^0| \geq 2d$, it suffices to prove this for a simple random walk S_t that moves $x \to x+1$ at rate λ and $x \to x-1$ at rate 1. To do this let $c = (\lambda-1)/2$ and observe that a now familiar large deviations estimate implies

(i) $$P_1(S_t < ct) \leq Ce^{-\gamma t},$$

where P_1 is the law of the process starting from $S_0 = 1$. If we let $T_0 = \inf\{ t : S_t \leq 0 \}$ then for $x > 0$

(ii) $$P_x (T_0 < \infty) = \lambda^{-x}.$$

(Since $\lambda^{-S(t)}$ is a martingale.) Combining the last two estimates shows that

$$P_1(S_u \leq 0 \text{ for some } u \geq t) \leq Ce^{-\gamma t} + \lambda^{-ct},$$

which implies the desired result.

Turning now to (b), we see that there is no point in checking this, since $\xi_\infty^1 = \mathbb{Z}^d$, so (c) is more difficult. To prove (c) we need to show:

(2) There are contants $C, c, \gamma \in (0,\infty)$ so that if $\| x \| < ct$, then

$$P(x \in \xi_t^0, \tau = \infty) \leq Ce^{-\gamma t}.$$

Proof: In Chapter 3 we went a long way toward proving this. We defined balls $R_k =$

$\{\,x : \|\,x\,\|_2 < \beta_k\,\}$, and times $s_k = \beta_k/\mu$, where $\beta_k = 2^k\beta$ and $\mu > 0$, and proved that

$$P(\,\xi_t^{R(k-1)} \supset R_k \text{ for all } t \in [s_k, s_k + s_{k+1}]\,) \geq 1 - Ce^{-\gamma\beta(k)}$$

where $C, \gamma \in (0,\infty)$ are independent of β. The last estimate shows that once R_0 is covered, a "chain reaction" starts which covers all the other balls with high probability.

In order to prove (2), we have to improve the last conclusion to show that the last construction always works when the process does not die out, and also that we do not have to wait too long to get the chain reaction started. Bramson and Griffeath (1980b) have a simple argument for doing this. They begin by observing that the proof above implies that if $\delta = (\lambda{-}1)/5$ (one possible choice for the constant δ in (5) in Chapter 3) and $\epsilon > 0$, there is a time s_0 (independent of A) so that

$$P(\text{ for some } y, \ B_{y,\delta s} \subset \xi_s^A \mid \tau^A < \infty\,) \geq 1{-}\epsilon.$$

To prove this, first get rid of the conditioning by observing that $P(\,\tau^A = \infty\,) \geq P(\,\tau^0 = \infty\,) > 0$, then look at the result in the last paragraph, and use a restart construction to find a time when the process contains R_0.

With the last result established, induction yields

$$P(\text{ for some } y \text{ and } t \leq ns, \ B_{y,\delta s} \subset \xi_t^A \mid \tau^A < \infty\,) \geq 1{-}\epsilon^n.$$

Setting s=an, and plugging the last result into (13) in Chapter 3 gives

$$\overline{P}(\text{ for some } y, \ B_{y,\delta t} \subset \xi_{t+an^2}^0 \text{ for all } t \geq 0\,) \geq 1 - \epsilon^n - Ce^{-\gamma an}.$$

The last detail is to get bounds on the location of y. Comparing with Richardson's model, we see that there are constants b,C, and γ (recall the last two constants change from line to line) so that $P(y \notin B_{0,bn^2}) \leq Cexp(-\gamma n^2)$, and it follows that

$$\overline{P}(B_{0,\delta t} \subset \xi^0_{t+(a+b/\delta)n^2} \text{ for all } t \geq 0) \geq 1 - Ce^{-\gamma n}.$$

Taking $t = n^2$ in the last result gives (2) with $Cexp(-\gamma t^{1/2})$ on the right–hand side. This is not the result advertised in (2), but is enough to prove the shape theorem using the argument in the last section. To see this, notice that: (i) the last result is II for $x = 0$; (ii) combining the argument for (2) above with the restart construction in the last section proves II for a general x; and (iii) once II is established, hypotheses (b) and (c) are not used in the proof. The last sentence is clearly a lot to swallow but is easy to "prove" — Bramson and Griffeath (1980b) prove their result using this approach. Alternatively, one can use the proof for the contact process which we are about to describe (briefly) to check hypothesis (c), but either way the remaining details are left to the reader.

Turning now to the contact process, we will focus on what is known and only rather briefly describe how these facts are proved. See Figure 11.5 for a simulation of the process with $\lambda = 1$ in $\mathbb{Z}^2$. In Durrett and Griffeath (1982) (abbr. DG(82)), it was shown that if $\lambda > \lambda_c(\mathbb{Z})$, the critical value for the contact process on $\mathbb{Z}$, then the conclusions of the shape theorem in the last section hold. The verification of (a) was done in Section 10b. The arguments for (b) and (c) are more involved. As the reader can probably guess, the proof involves using imbedded one dimensional processes, but the proof gets messy because if we pick a particle, then the contact process it generates survives forever with probability $\rho < 1$; and if it dies out, it effects in a negative way the survival of nearby processes.

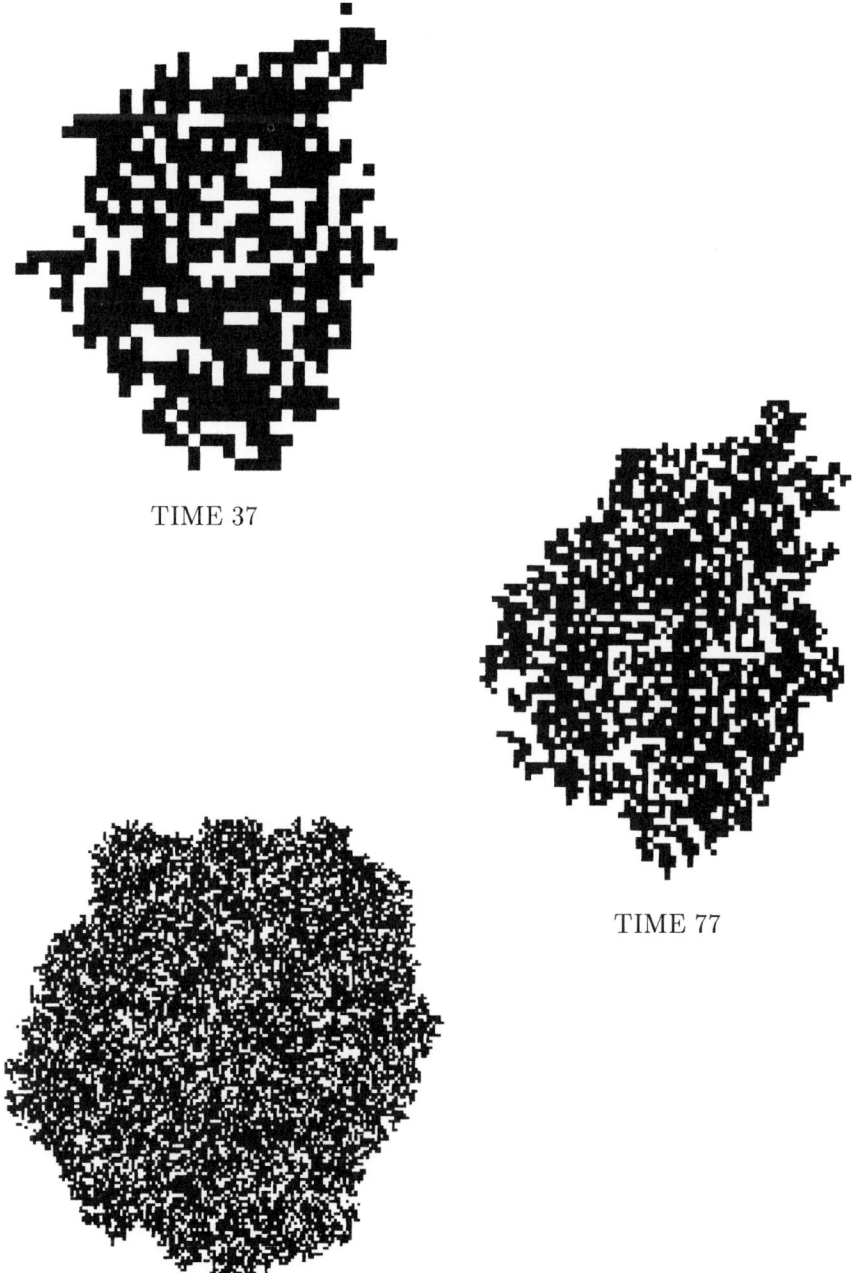

TIME 37

TIME 77

TIME 162   Figure 11.5

The negative dependence referred to in the last paragraph is neutralized by looking at the contact process on $\mathbb{Z}^+ = \{\ 0,\ 1,\ 2,\ ...\ \}$, but this forces us to show that $\lambda_c(\mathbb{Z}^+) = \lambda_c(\mathbb{Z})$ (exercise for the reader, use the renormalized bond construction), and that the exponential estimates in Section 10b hold for the process on $\mathbb{Z}^+$. All of this adds up to a lot of work to be done, so instead of boring the reader (and the author) with the unpleasant details, I will leave you with the problem of figuring out how to show that

$$P(\ t(x) > t,\ \tau = \infty\ ) \le Ce^{-\gamma t}\ \text{ for } \|x\| < ct.$$

The proof of the last result is a little like driving from Los Angeles to New York. It takes a little thought to plan the trip, and then a lot of patience to carry it out. If you get stuck or run out of gas, an answer can be found on pages 546–549 of DG(82). If you find a better way of doing this, let me know.

The result of DG(82) is almost the last word on shape theorems for the contact process. Durrett and Schonmann (1987) (hereafter DS(87)) have proved a result that includes the case of a "symmetric discrete time finite range contact process in $\mathbb{Z}^{2}$"; that is, a process constructed from a graphical representation in which bonds $(x,n) \to (x+y,n+1)$ $y \in S$ are independently open with probability p, and S is a finite set with S $= -S$. In this case they show that if $p > p_c(\ \mathbb{Z} \times \{-L,...,L\}\ )$, the critical value for the process restricted to $\mathbb{Z} \times \{-L,...,L\}$, then the shape theorem holds. We conjecture

(3) $p_c(\mathbb{Z} \times \{-L,...L\}) \downarrow p_c(\mathbb{Z}^2)$

and it is an important open problem to prove that (3) holds.

Durrett and Schonmann are able to prove results for $p > p_c(\mathbb{Z} \times \{-L,...L\})$ because they can prove results for finite range systems in one dimension. The key to doing that is to observe that (i) the existence of edge speeds and the exponential estimate

$$(4) \qquad\qquad P(\, r_n > at\,) \leq Ce^{-\gamma n} \ \text{for a} > \alpha(p),$$

are valid for finite range systems on $\mathbb{Z}$ or on $\mathbb{Z} \times \{-L,...L\}$, and (ii) for a symmetric system, (4) is all we need to show that the probability of having paths in the parallelograms of Section 10a approaches 1 as $K \to \infty$.

With (ii) in hand, we can try the renormalized bond construction, but there is one problem; in the finite range case, paths can jump over each other without hitting. To circumvent this difficulty, we observe that two paths that cross will hit with a probability $\geq \epsilon > 0$ (independent of L), and then give ourselves lots of opportunities to make the desired connections. Here a picture is worth (and replaces) a thousand words.

Figure 11.6

See Figure 11.6. The large tubes are the renormalized bonds and the short tubes are connectors used to tie the long tubes together.

The details of the proof are not difficult, but are a little messy and are left to the reader's imagination. (See DS(87) for details.) Once the renormalized bond construction is developed, the rest is easy. The construction gives us positive probability of percolation so $p_c \leq \inf\{ p : \alpha(p) > 0 \}$, and the arguments in Section 4a take over to give us

$$p_c = \inf\{ p : \alpha(p) > 0 \} = \sup\{ p : \alpha(p) < 0 \}.$$

The complete convergence theorem

$$\xi_t^A \Rightarrow P(\tau^A < \infty) \, \delta_\phi + P(\tau^A = \infty) \, \xi_\infty^1$$

can be proved by using an idea of David Griffeath: "it suffices to show that if ξ_t^A and ξ_t^B are independent,

$$P(\xi_t^A \cap \xi_t^B \neq \phi, \; \tau^A \geq t \; \tau^B \geq t) \rightarrow 0."$$

(Observe that $\{ \xi_{2t}^A \cap B \neq \phi \} = \{ \xi_t^A \cap \check{\xi}_t^B \neq \phi \}$.), To prove the result in quotation marks, we use the renormalized bond construction to make sure that the two processes hit each other when they don't die out. Finally, the arguments of Section 11b can be repeated almost verbatim to prove exponential estimates and the strong law for $|\xi_t^0|/t$. Again, for details see DS(87).

Before the reader gets lulled into thinking that everything is easy, we would like to point out that we do not know how to prove things for the contact process on

$\mathbb{Z} \times \{-L,..L\}$. The phrase that causes trouble is "two paths that cross will hit with a probability $\geq \epsilon > 0$ (independent of L)." In the discrete time finite range situation, there is a strict upper bound on the speed of paths, but this is not true in continuous time and we do not know how to rule out the possibility that some malicious demon causes the paths we have chosen to move extremely fast (with a speed that increases with L) just at the moment they are about to cross. The last fear is clearly a paranoid delusion, but we do not know how to rule it out. This is clearly a technical problem and not as much fun as thinking about (3), but its solution would be a useful addition to the subject.

Finally, we promised the reader some "applications" to bring the book to a climax. Two situations in which contact process type models might be useful are in the studies of raccoon rabies and chestnut blight. The first situation is one of some concern as this is being written. An epidemic has been spreading north and east from the Virginia–West Virginia border since 1977 and is now approaching the New York border. The second situation is much older news. In 1904 the fungus *Endothia parasitica* was discovered at the Bronx Zoological Park in New York City, and by 1945 the blight had spread across the tree's natural range in the eastern United States, killing almost all of the chestnut trees in the process.

It is amusing to note that the articles describing these epidemics (in the *Ithaca Journal* Feb. 19, 1987 and July 24, 1986, see also the December 1986 *Discover* magazine) say raccoon rabies is "moving toward New Jersey and New York at an estimated speed of 25–30 miles per year," and "moving at 10 miles per year, the [chestnut] blight swept south." (They say nothing, however, about these epidemics having an asymptotic shape.) Like many applications, these two epidemics point out inadequacies in the current results, and point the way for future research. As the reader has probably noticed, we said above that the rabies epidemic has been spreading

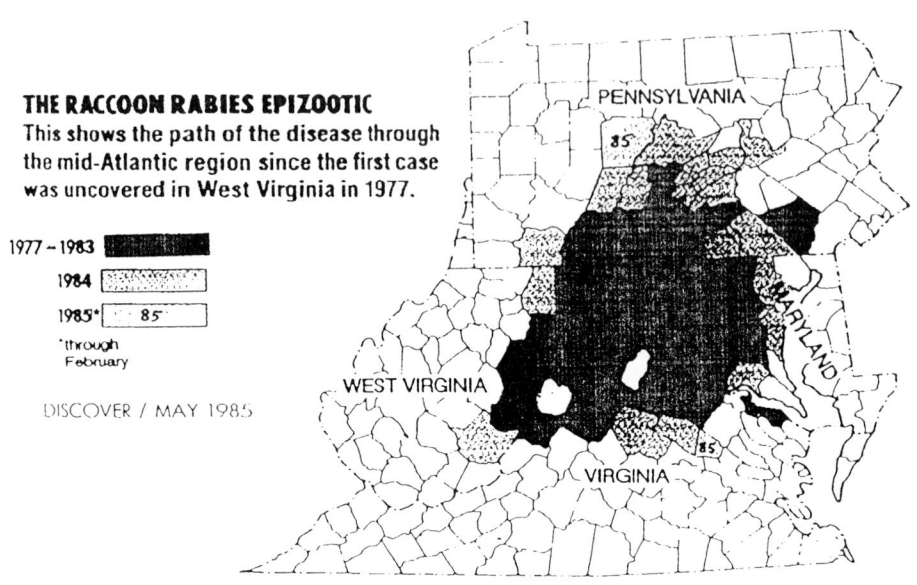

THE RACCOON RABIES EPIZOOTIC
This shows the path of the disease through
the mid-Atlantic region since the first case
was uncovered in West Virginia in 1977.

1977 – 1983
1984
1985*   85
*through
February

DISCOVER / MAY 1985

Figure 11.7

north and east but not south and west. (See Figure 11.7.) This undoubtedly has

something to do with the fact that the United States is not homogeneous like $\mathbb{Z}^2$. One

solution is to change the model to a contact process in a random environment (CPRE);

that is, the birth rate is λ times the number of occupied neighbors, but the death rate

at a site is $\delta(x)$, where $\delta(x)$ $x \in \mathbb{Z}^2$ is a stationary process. In formulating the last

model, we are thinking that the animal (and hence its birth rate) stays the same, but

the hospitality of the environment changes. The problem for the CPRE is to show that

when the infection does not die out, it grows linearly and has an asymptotic shape.

Before the reader dismisses this as trivial, she should remember that RWRE's have

drastically different behavior from random walks. We conjecture that the statement

about linear growth is true in $d \geq 2$, but not in $d = 1$.

Turning to chestnut blight, our inspiration comes from the observation that "In

1910, a Pennsylvania blight commission ordered that a 10 mile wide swath be cut to isolate the diseased trees, but airborne fungus leapfrogged over the quarantined area." It is frightening to think of fungus flying more than 10 miles through the air, but the last quote indicates that we are dealing with a process with a long range interaction (or perhaps that the people who were cutting down the trees got the fungus on their boots).

Bramson, Durrett, and Swindle (198?) have recently studied what happen when the range of the interaction in the contact process $R \to \infty$. They showed that if we change the parameter to $\beta =$ the total birth rate due to an isolated particle, then $\beta_c(R) \to 1$, and the probability of the epidemic surviving for all time starting from one infected individual approaches $(\beta-1)/\beta$, the answers for a branching process. The main result of the paper identifies the rate at which in $\beta_c(R) \to 1$:

$$\beta_c(R) - 1 \approx \begin{cases} C \: / \: R^{2/3} & \text{in } d = 1 \\ C \: (\log R)/R^2 & \text{in } d = 2 \\ C \: / \: R^d & \text{in } d \geq 3. \end{cases}$$

Results in Swindle's (1988) thesis give detailed information about how the contact process spreads when R is large, the answers being in terms of a nonlinear partial differential equation and a generalized Ornstein–Uhlenbeck process. Some indication of these results can be found in Durrett's two (1988) papers on "Crabgrass, measles, and gypsy moths." Crabgrass is the $R \to \infty$ limit.

NOTES

In this section we pay homage to the past and future. We briefly describe the history of the results in the book and we indicate where the reader can find out about recent developments. Numbers at the beginning of each paragraph indicate the chapter or section to which the remarks refer. The last paragraph lists some recent conference proceedings.

1. Eden (1961) introduced a discrete time model in which at each time a randomly chosen boundary site is added to the set. Richardson (1973) was able to analyze this model and a number of others by considering the system in continuous time. Our treatment here is based on extensions of his work due to Kingman (1973), Hammersley (1974), Durrett (1980), and Liggett (1985a). Believe it or not there are still a number of open problems concerning Richardson's model; see Racz and Plischke (1985), Leyvraz (1985), and Kardar, Parisi, and Zhang (1986).

2. Holley and Liggett (1975) proved the basic results about the voter model given in Chapter 2. The model is much older. It appeared in a paper of Clifford and Sudbury (1973) under the name "invasion process," and as early as 1943 in the genetics literature under the name "stepping stone model." For an account of the history, see Cox and Griffeath (1987). We will have more to say about the voter model in the notes on 10 below.

3. Williams and Bjerknes (1972) invented the biased voter model. They simulated the model with $\lambda = 1.1$, and thought it was a fractal. That was disproved by Mollison (1972). Bramson and Griffeath (1980b,1981) proved the shape theorem, which is the main result of the chapter.

4. Harris (1974) invented the contact process. For an account of the early work on this process, see Griffeath (1981).

5a. Oriented percolation in two dimensions is surveyed in Durrett (1984a).

5b. This section was inspired by a picture in Kinzel (1985). The main results are from Durrett and Griffeath (1983). Here we have assumed spatial symmetry. For an account of what can happen when this assumption is dropped, see Schonmann (1986). For more on PCA's (probabilistic cellular automata) from the physicist's viewpoint, see Grinstein, Jayaprakash, and He (1985), and Bennet and Grinstein (1985). The second paper contains a very interesting model.

5c. Graphical representations were invented by T. Harris (1978), and developed in Griffeath's (1979) monograph.

5d. Pascal's triangle mod 2 is one of Wolfram's favorite processes and has appeared in a number of his papers. The main result was proved by Miyamoto (1979), and later independently by Lind (1984).

5e. Cancellative duality was invented by Holley and Stroock (1979) but some credit should go to Matloff (1977). The graphical viewpoint is due to Griffeath (1979).

6. The main result of Section 6a is due to Kesten (1980a). Our treatment gives Russo a little more credit than he deserves. The key ideas about sponge crossings were hatched by Russo (1978) and Seymour and Welsh (1978) independently. Kesten's (1982) book and Wierman's (1982) article can be consulted for more about the subject. For an interesting "Approximate 0–1 law" related to the developments at the end of Section 6a, see Russo (1982). Recent developments are explained in Kesten (1987b). Two topics that deserve special mention are the random surfaces of Aizenman, Chayes, Chayes, Fröhlich, and Russo (1983), and invasion percolation of Chayes, Chayes, and Newman (1985).

7. The process under consideration was invented by Mandelbrot (1974). See also

his (1983) book. The results proved here are from Chayes, Chayes, and Durrett (1988).

8a. Hammersly and Welsh (1965) invented first passage percolation. For an account of early developments, see Smythe and Wierman (1978). Soon after their book came out, there was an explosion of results weakening the hypotheses on results concerning the time constant (Reh (1979), Wierman (1980), Branvall (1980), and Cox and Durrett (1981)) and sharpening other results (Kesten (1980b)). Grimmett and Kesten (1984) have an interesting account of the connection with network flows and electrical resistances. A more recent survey is Kesten (1986). Kesten (1987a) discusses a related problem concerning surfaces with random weights for which there are still a number of open problems.

8b. This material is from Chayes, Chayes, and Durrett (1986). At several points we avoid using an inequality due to van den Berg and Kesten (1985) which is the opposite of Harris' inequality. The probability two increasing events "occur disjointly" $A \circ B$ satisfies $P(A \circ B) \leq P(A)P(B)$. It is not known whether the last inequality holds for ANY two events, see van den Berg and Fiebig (1987). An interesting variation of the system studied in 8b is the model in which bonds have travel times 1 and ∞ with probabilities p and 1$-$p. This time one wants to study how the time constant diverges as $p \downarrow 1/2$. See Ritzenberg and Cohen (1984), Edwards and Kerstein (1985), and M. Barma (1985).

9. The model considered here is called the "spatial epidemic with removal." For early results see Bailey (1965) and Mollison (1977). The main result of this section is due to Cox and Durrett (1988). For the physicist's viewpoint, see Grassberger (1983), McKay and Jan (1984), Cardy and Grassberger (1985), von Niessen and Blumen (1986), Ohtsuki and Keyes (1986), and Peleti (1986).

10. The results in 10a are from Cox and Griffeath (1986); those in 10b are from Cox (198?). One could write a book about the voter model. Two important early

results of Bramson and Griffeath, which should have been treated in detail in Liggett's book and are also ignored here, are their (1979) study of the stationary distributions in $d \geq 3$, and their (1980a) paper on the rate at which the density of particles in coalescing random walks goes to 0. We should also mention recent work on occupation times (central limit theorems and large deviations) for the voter model and other systems (independent and branching random walks). References to this work and other recent developments can be found in Cox and Griffeath (1986) and (1987).

11. The renormalized bond construction was invented by Durrett and Griffeath (1983), who used it to prove the results in 11b. The general shape theorem explained in 11c was formulated by Durrett and Griffeath (1982), and used to prove the result for the contact process described in 11d. At the end of 11d, we mentioned some recent work of Bramson, Durrett, and Swindle (198?). Some other things that we could have mentioned are: (i) asymptotics for the critical value as the dimension $\rightarrow \infty$: Holley and Liggett (1981), Griffeath (1983), and Schonmann and Vares (1986); and (ii) behavior on a finite set: Schonmann (1985), Durrett and Liu (1988), Durrett and Schonmann (1988).

General References: Durrett (1984b), Tautu (1986), and Kesten (1987c) are conference proceedings which can be consulted for recent developments. If you want to break new ground, Farmer, Toffoli, and Wolfram (1984); Demongeot, Golès, and Tchuente (1985); and Stanley and Ostrowsky (1986) are collections of nonrigorous articles which should provide food for thought.

REFERENCES

We have listed only the papers we referred to. For more references, see Liggett (1985b), Kesten (1986,1987b), and the sources mentioned in the NOTES.

M. Aizenman, J.T. Chayes, L. Chayes, J. Fröhlich, and L. Russo (1983) On a sharp transition from area to perimeter law in a system of random surfaces. *Comm. Math. Phys.* 92, 19–69.

D. Amati, G. Marchesini, N. Ciafaloni, and G. Parisi (1976) Expanding disc as dynamic vacuum instability in Reggeon field theory. *Nucl. Phys. B* 114, 483–504.

R. Arratia (1981) Limiting point processes for rescalings of coalescing and annihi-lating random walks on $\mathbb{Z}^d$. *Ann. Prob.* 9, 909–936.

K.B. Athreya and P.E. Ney (1972) *Branching Processes.* Springer Verlag, New York.

_____ (1978) A new approach to the limit theory of recurrent Markov chains. *Trans AMS* 245, 493–501.

N.T.J. Bailey (1965) The simulation of stochastic epidemics in two dimensions. In Vol. IV of *Proc. 5th Berkeley Symp.*, U. of California Press.

M. Barma (1985) Shortest paths in percolation. *J. Phys. A* 18, L277–283.

C. Bennett and G. Grinstein (1985) The role of irreversibility in stabilizing complex and nonergodic behavior in locally interacting discrete systems. *Phys. Rev. Letters* 55, 657–660.

J. van den Berg and U. Fiebig (1987) On a combinatorial conjecture concerning disjoint occurrences of events. Ann. Prob. 15, 354–374.

J. van den Berg and H. Kesten (1985) Inequalities with applications to percolation and reliability. *J. Appl. Prob.* 22, 556–569.

J.D. Biggins (1978) The asymptotic shape of branching random walk. *Adv. Appl. Prob.* 10, 62–84.

J. Blease (1977a) Series expansions for the directed–bond percolation problem. *J. Phys. C* 10, 917–924.

_____ (1977b) Pair–connectedness for directed bond percolation on some two dimensional lattices by series methods. *J. Phys. C.* 10, 3461–3476

M. Bramson, R. Durrett, and G. Swindle (198?) Statistical mechanics of crabgrass. Submitted to *Ann. Prob.*

M. Bramson and L. Gray (1985) The survival of branching annihilating random walk. *Z. fur Wahr.* 68, 447–460.

M. Bramson and D. Griffeath (1979) Renormalizing the three dimensional voter model. *Ann. Prob.* 4, 418–432.

———— (1980a) Asymptotics for interacting particle systems on $\mathbb{Z}^d$. *Z. fur Wahr.* 45, 183–196.

———— (1980b) On the Williams–Bjerknes tumor growth model, II. *Proc. Camb. Phil. Soc.* 88, 339–357.

———— (1981) On the Williams–Bjerknes tumor growth model, I. *Ann. Prob.* 9, 173–185.

G. Branvall (1980) A note on limit theorems in percolation. *Z. fur Wahr.* 53, 317–321.

S.R. Broadbent and J.M. Hammersley (1957) Percolation processes. *Proc. Camb. Phil. Soc.* 53, 629–645.

R.C. Brower, M.A. Furman, and M. Moshe (1978) Critical exponents for the Reggeon quantum spin model. *Physics Letters* 76 B, 2113–219.

J.L. Cardy and P. Grassberger (1985) Epidemic models and percolation *J. Phys. A* 18, L267–271.

J.L. Cardy and R.L. Sugar (1980) Directed percolation and Reggeon field theory. *J. Phys. A* 13, L423–427.

J.T. Chayes and L. Chayes (1986) An inequality for the infinite cluster density in percolation. *Phys. Rev. Lettters* 56, 1619–1623.

J.T. Chayes, L. Chayes, and R. Durrett (1986) Critical behavior of two dimensional first passage percolation. *J. Stat. Phys.* 45, 933–951.

———— (1988) Connectivity properties of Mandelbrot's percolation process. *Prob. Th. Rel. Fields*, to appear

J.T. Chayes, L. Chayes, D.S. Fisher, and T. Spencer (1986) Finite size scaling and correlation lengths for disordered systems. *Phys. Rev. Letters* 57, 2999–3002.

J.T. Chayes, L. Chayes, and C. Newman (1985) The stochastic geometry of invasion percolation. *Comm. Math. Phys.* 101, 383–407.

K.L. Chung (1974) *A Course in Probability Theory*, second edition. Academic Press, New York.

P. Clifford and A. Sudbury (1973) A model for spatial conflict. *Biometrika* 60, 581–588.

J.T. Cox (1988) Coalescing random walks and voter model consensus times on the torus in $\mathbb{Z}^d$. *Ann. Prob.*, to appear.

J.T. Cox and R. Durrett (1981) Some limit theorems for percolation processes with necessary and sufficient conditions. *Ann. Prob.* 9, 583–603.

———— (1988) Limit theorems for the spread of epidemics and forest fires. *Stoch. Proc. Appl.*, to appear.

J.T. Cox and D. Griffeath (1986) Diffusive clustering in the two dimensional voter model. *Ann. Prob.* 14, 347–370.

———— (1987) Recent results on the stepping stone model. In Kesten (1987c).

J. Demongeot, E. Golès, and M. Tchuente, editors (1985) *Dynamical Systems and Cellular Automata.* Academic Press, New York.

Y. Derriennic (1983) Un théoreme ergodique presque sous–additif. *Ann. Prob.* 11, 669–677.

D. Dhar and M. Barma (1981) Monte Carlo simulation of directed percolation on a square lattice. *J. Phys. C* 14, L1–6.

Z.V. Djordjevic, H.E. Stanley, and A. Margolina (1982) Site percolation threshold for honeycomb and square lattices. *J. Phys. A* 15, L405–412.

E. Domany and W. Kinzel (1981) Directed percolation in two dimensions: numerical analysis and an exact solution. *Phys. Rev. Letters* 47, 5–8.

N. Dunford (1951) An individual ergodic theorem for noncommutative transfor–mations. *Acta Sci. Math (Szeged)* 14, 1–14.

R. Durrett (1979) An infinite particle system with additive interactions. *Adv. Appl. Prob.* 11, 355–383.

———— (1980) On the growth of one dimensional contact processes *Ann. Prob.* 8, 890–907.

———— (1984a) Oriented percolation in two dimensions. *Ann. Prob.* 12, 999–1040.

————, editor (1984b) *Particle Systems, Random Media, and Large Deviations.* AMS Contemporary Math. Series, Vol. 41., American Math. Society, Providence, R.I.

———— (1988a) Crabgrass, measles, and gypsy moths: an introduction to inter–acting particle systems. *Math. Intelligencer*, to appear.

———— (1988b) Crabgrass, measles, and gypsy moths: an introduction to modern probability. *Bull. AMS*, to appear.

R. Durrett and L. Gray (1986) Some peculiar properties of a particle system with sexual reproduction. In Tautu (1986).

_____ (198?) Some peculiar properties of a particle system with sexual repro–duction. Submitted to *Ann. Prob.*

R. Durrett and D. Griffeath (1982) Contact processes in several dimensions. *Z. fur Wahr.* 59, 535–552.

_____ (1983) Supercritical contact processes on $\mathbb{Z}$. *Ann. Prob.* 11, 1–15.

R. Durrett and T.M. Liggett (1981) The shape of the limit set in Richardson's growth model. *Ann. Prob.* 9, 186–193.

R. Durrett and Xiu–fang Liu (1988) The contact process on a finite set. *Ann. Prob.*, to appear.

R. Durrett and R.H. Schonmann (1987) Stochastic growth models. In H. Kesten (1987c).

_____ (1988) The contact process on a finite set, II. *Ann. Prob.*, to appear.

M. Eden (1961) A two dimensional growth process. In Vol. IV of *Proc. 4th Berkeley Symp.*, U. of California Press.

B.F. Edwards and A.R. Kerstein (1985) Is there a lower critical dimension for chemical distance? *J. Phys. A* 18, L1081–1086.

J.W. Essam and De'Bell (1981) Series expansion studies of directed percolation: estimates of correlation length exponents. *J. Phys. A* 14, L459–461.

D. Farmer, T. Toffoli, and S. Wolfram, editors (1984) *Cellular Automata.* (*Physica* 10 D, 1–245) Published separately by North Holland, Amsterdam.

H.L. Frisch and J.M. Hammersley (1963) Percolation processes and related topics. *SIAM Journal* 11, 894–918.

A. Galves and E. Presutti (1987) Edge fluctuations for the one dimensional contact process. *Ann. Prob.* 15, 1131–1145.

P. Grassberger (1982) On the phase transitions in Schlögl's second model. *Z. Phys. B* 47, 365–374.

_____ (1983) On the critical behavior of the general epidemic process and dynam–ical percolation. *Math. Biosci.* 63, 157–172.

_____ (1985) On the spreading of two dimensional percolation. *J. Phys. A* 18, L215–219.

P. Grassberger and A. de la Torre (1979) Reggeon field theory (Schlögl's second model) on a lattice: Monte Carlo calculations of critical behavior. *Ann. Phys.* 122, 373–396.

P. Grassberger and K. Sundermeyer (1978) Reggeon field theory and Markov processes. *Physics Letters* 77 B, 220–222.

L. Gray (1982) The positive rates problem for attractive nearest neighbor systems on $\mathbb{Z}$. *Z. fur Wahr.* 61, 389–404.

_____ (1986) Duality for general spin systems, with applications in one dimension. *Ann. Prob.* 14, 371–396.

D. Griffeath (1979) *Additive and Cancellative Interacting Particle Systems.* Lecture Notes in Math 724, Springer Verlag, New York.

_____ (1981) The basic contact process. *Stoch. Proc. Appl.* 11, 151–185.

_____ (1983) The binary contact path process. *Ann. Prob.* 11, 692–705.

G.R. Grimmett and H. Kesten (1984) First passage percolation, network flows, and electrical resistances. *Z. fur Wahr.* 66, 335–366.

G. Grinstein, C. Jayaprakash, and Y. He (1985) Statistical mechanics of probabilistic cellular automata. *Phys. Rev. Letters* 55, 2527–2530.

J.M. Hammersley (1961) Comparison of atom and bond percolation processes. *J. Math. Phys.* 2, 728–733.

_____ (1974) Postulates for subadditive processes. *Ann. Prob.* 2, 652–680.

J.M. Hammersley and D.J.A. Welsh (1965) First passage percolation, subadditive processes, stochastic networks, and generalized renewal theory. In *Bernoulli Bayes Laplace Anniversary Volume*, ed. by J. Neyman and L. LeCam, Springer Verlag, New York.

T.E. Harris (1960) A lower bound for the critical probability in a certain percolation process. *Proc. Camb. Phil. Soc.* 56, 13–20.

_____ (1974) Contact interactions on a lattice. *Ann. Prob.* 2, 969–988.

_____ (1978) Additive set–valued Markov processes and graphical methods. *Ann. Prob.* 6, 355–378.

R. Holley (1972) Markovian interaction processes with finite range interactions. *Ann. Math. Stat.* 43, 1961–1967.

R. Holley and T. Liggett (1975) Ergodic theorems for weakly interacting systems and the voter model. *Ann. Prob.* 3, 643–663.

_____ (1978) The survival of contact processes. *Ann. Prob.* 6, 198–206.

_____ (1981) Generalized potlatch and smoothing processes. *Z. fur Wahr.* 55, 165–195.

R. Holley and D. Stroock (1979) Dual processes and their applications to infinite interacting systems. *Adv. Math.* 32, 149–174.

J.P. Kahane and J. Peyrière (1976) Sur certaines martingales de B. Mandelbrot. *Advances in Math.* 22, 131–145.

M. Kardar, G. Parisi, and Y.C. Zhang (1986) Dynamical scaling of growing inter-faces. *Phys. Rev. Letters* 56, 889–892.

J. Kertesz and T. Vicsek (1980) Orientated [sic] bond percolation. *J. Phys. C* 13, L343–348.

H. Kesten (1980a) The critical probability of bond percolation on the square lattice equals 1/2. *Comm. Math. Phys.* 74, 41–59.

_____ (1980b) On the time constant and path length in first passage percolation. *Adv. Appl. Prob.* 12, 848–863.

_____ (1982) *Percolation Theory for Mathematicians.* Birkhauser, Boston.

_____ (1986) Aspects of first passage percolation. In Lecture Notes In Math 1180, Springer Verlag, New York.

_____ (1987a) Surfaces with minimal random weights and maximal flows: a higher dimensional version of first passage percolation. *Ill. J. Math.* 31, 99–166.

_____ (1987b) Percolation theory and first passage percolation. *Ann. Prob.* 15, 1231–1271

_____ , editor (1987c) *Percolation Theory and the Ergodic Theory of Interacting Particle Systems.* Vol. 8 in the I.M.A. Volumes in Math. and its Appl., Springer Verlag, New York.

_____ (1987d) Scaling relations for 2D–percolation. *Comm. Math. Phys.* 109, 109–156.

J.F.C. Kingman (1973) Subadditive ergodic theory. *Ann. Prob.* 1, 883–909.

W. Kinzel (1983) Directed percolation. In *Percolation Structures and Processes*, ed. by G. Deutscher, R. Zallen, and J. Adler, Adam Higgins Pub. Co., Bristol, England.

_____ (1985) Phase transitions in cellular automata. *Z. Phys. B* 58, 229–244.

W. Kinzel and J. Yeomans (1981) Directed percolation: A finite size scaling appraoch. *J. Phys. A* 14, L163–168.

F. B. Knight (1981) *Essentials of Brownian Motion and Diffusion.* AMS Math Survey 18, American Math. Society, Providence, R.I.

K. Kuulasmaa (1982) The spatial general epidemic model and locally dependent random graphs. *J. Appl. Prob.* 19, 745–758.

K. Kuulasmaa and S. Zachary (1984) On spatial general epidemics and bond perco—lation processes. *J. Appl. Prob.* 21, 911–914.

F. Leyvraz (1985) The "active perimeter" in cluster growth models: a rigorous bound. *J. Phys. A* 18, L941–945.

T.M. Liggett (1985a) An improved subadditive ergodic theorem. *Ann. Prob.* 13, 1279–1285.

———— (1985b) *Interacting Particle Systems*. Springer Verlag, New York.

D. Lind (1984) Applications of ergodic theory and sofic systems to cellular automata. *Physica 10 D*, 36–44.

C. McDiarmid (1980) Clutter percolation and random graphs. In *Math. Programming Study* 13, North Holland, Amsterdam.

———— (1981) General percolation and random graphs. *Adv. Appl. Prob.* 13, 40–60.

G. McKay and N. Jan (1984) Forest fires as critical phenomena. *J. Phys. A* 17, L757–760.

B. Mandelbrot (1974) Intermittent turbulence in self—similar cascades: Divergence of high moments and dimension of the carrier. *J. Fluid. Mech.* 62, 331–358.

———— (1983) *The Fractal Geometry of Nature*. W.H. Freeman and Co., New York.

O. Martin, A.M. Odlyzko, and S. Wolfram (1984) Algebraic properties of cellular automata. *Comm. Math. Phys.* 93, 219–258.

N. Matloff (1977) Ergodicity conditions for a dissonant voting model. *Ann. Prob.* 5, 371–386.

———— (1980) A dissonant voting model, II. *Z. fur Wahr.* 51, 63–78.

R.D. Mauldin, S. Graf, and S.C. Williams (1987) Exact Hausdorff dimension in random recursive constructions. *Proc. Nat. Acad. Sci.* 84, 3959–3961.

R.D. Mauldin and S.C. Williams (1986) Random recursive constructions: Asymptotic geometric and topological properties. *Trans. AMS* 295, 325–346.

M. Miyamoto (1979) An equilibrium state for a one dimensional Life game. *J. Math. Kyoto Univ.* 19, 525–540.

D. Mollison (1972) Conjecture on the spread of an infection in two dimensions disproved. *Nature* 240, 467–468.

———— (1977) Spatial contact models for ecological and epidemic spread. *J. Roy. Stat. Soc. B* 39, 283–326.

———— (1978) Markovian contact processes. *Adv. Appl. Prob.* 10, 85–108.

C.M. Newman and A.L. Wright (1981) An invariance principle for certain dependent sequences. *Ann. Prob.* 9, 671–675.

W. von Niessen and A. Blumen (1986) Dynamics of forest fires as a directed percolation model. *J. Phys. A* 19, L289–293.

T. Ohtsuki and T. Keyes (1986) Biased percolation: Forest fires with wind. *J. Phys. A* 19, L281–287.

N. Packard and S. Wolfram (1985) Two dimensional cellular automata. *J. Stat. Phys.* 38, 901–946.

L. Peleti (1986) Field theories of walks and epidemics. In Stanley and Ostrowsky (1986).

L. Pietronero and E. Tosati (1986) *Fractals in Physics.* North Holland, Amsterdam.

E. Presutti and H. Spohn (1983) Hydrodynamics of the voter model. *Ann. Prob.* 11, 867–875.

Z. Racz and M. Plischke (1985) Active zone of growing clusters: Diffusion limited aggregation and the Eden model in two and three dimensions. *Phys. Rev. A* 31, 985–994.

W. Reh (1979) First passage percolation under weak moment conditions. *J. Appl. Prob.* 16, 750–763.

D. Richardson (1973) Random growth in a tesselation. *Proc. Camb. Phil. Soc.* 74, 515–528.

A.L. Ritzenberg and R.J. Cohen (1984) First passage percolation: Scaling and critical exponents. *Phys. Rev. B* 30, 4038–4040.

L. Russo (1978) A note on percolation. *Z. fur Wahr.* 43, 39–48.

———— (1981) On the critical percolation probabilities. *Z. fur Wahr.* 56, 229–237.

———— (1982) An approximate 0–1 law. *Z. fur Wahr.* 61, 129–139.

R.H. Schonmann (1985) Metastability for the contact process. *J. Stat. Phys.* 41, 445–464.

———— (1986) The asymmetric contact process. *J. Stat. Phys.* 44, 505–534.

R.H. Schonmann and M.E. Vares (1986) The survival of the large dimensional basic contact process. *Prob. Th. Rel. Fields* 72, 387–393.

K. Schürger (1979) On the asymptotic behavior of a class of contact interaction processes. *Z. fur Wahr.* 48, 35–48.

———— (1980) On the asymptotic behavior of percolation processes. *J. Appl. Prob.* 17, 385–402.

P.D. Seymour and D.J.A. Welsh (1978) Percolation probabilities on the square lattice. *Ann. Discrete Math.* 3, 227–245.

R. Smythe (1976) Multiparameter subadditive processes. *Ann. Prob.* 4, 772–782.

R. Smythe and J.C. Wierman (1978) *First Passage Percolation on the Square Lattice.* Lecture Notes In Math 671, Springer Verlag, New York.

F. Spitzer (1976) *Principles Of Random Walk,* second edition. Springer Verlag, New York.

B. Spock and M.B. Rothenberg (1985) *Dr. Spock's Baby and Child Care,* 40th Anniversary Edition. Simon and Schuster, New York.

H.E. Stanley and N. Ostrowsky, editors (1986) *On Growth and Form.* Martinus Nijhoff, Boston.

G. Swindle (1988) Hydrodynamic limits for the contact process with large range. Ph.D. thesis, Cornell University.

M.F. Sykes and J. F. Essam (1964) Exact critical probabilities for site and bond problems in two dimensions. *J. Math. Phys.* 5, 117–1127.

P. Tautu, editor (1986) *Stochastic spatial processes.* Lecture Notes In Math 1212, Springer Verlag, New York.

S. Tavaré (1984) Line of descent and genealogical processes, and their applications in population genetics models. *Theor. Pop. Biol.* 26, 119–164.

G.Y. Vichniac, P. Tamayo, and H. Hartman (1986) Annealed and quenched inhomo−geneous cellular automata. *J. Stat. Phys.* 45, 875–883.

J.C. Wierman (1980) Weak moment conditions for time coordinates in first passage percolation models. *J. Appl. Prob.* 17, 968–978.

J.C. Wierman (1982) Percolation theory. *Ann. Prob.* 10, 509–524.

J.C. Wierman and W. Reh (1978) On conjectures in first passage percolation theory. *Ann. Prob.* 6, 388–397.

T. Williams and R. Bjerknes (1972) Stochastic models for abnormal clone spread through epithelial basal layer. *Nature* 236, 19–21.

S. Wolfram (1983) Statistical mechanics of cellular automata. *Rev. Mod. Phys.* 51, 601–644.

———— (1984) Universality and complexity in cellular automata. *Physica* 10 D, 1–35.

Y. Zhang and Y.C. Zhang (1984) A limit theorem for $N(0,n)/n$ in first passage perco−lation. *Ann. Prob.* 12, 1068–1076

H. Ziezold and C. Grillenberger (1985) On the critical infection rate of the one dimensional basic contact process: Numerical results. Preprint.

INDEX